Field Guide to the Palms of New Guinea

William J. Baker and John Dransfield

Line drawings by Patricia K. R. Davies,
Soejatmi Dransfield and Lucy T. Smith

Kew Publishing
Royal Botanic Gardens, Kew

PLANTS PEOPLE
POSSIBILITIES

First published in 2006 by
Royal Botanic Gardens, Kew
Richmond, Surrey, TW9 3AB, UK
www.kew.org

ISBN 1 84246 138 9

British Library Cataloguing in Publication Data
A catalogue record for this book is available from the British Library

Production Editor: Michelle Payne
Typesetting and page layout: Margaret Newman
Design by Media Resources, Information Services Department,
Royal Botanic Gardens, Kew

Printed and bound in Italy by Printer Trento

For information or to purchase all Kew titles please visit
www.kewbooks.com or email publishing@kew.org

All proceeds go to support Kew's work in saving the world's plants for life

Cover photo: *Cyrtostachys loriae*, Timika, Papua (Photo: W. J. Baker)

Table of contents

Acknowledgements

The book has been produced as a result of the UK Darwin Initiative Papuan Plant Diversity Project (DIPPDP), a collaborative capacity building initiative based at the Royal Botanic Gardens, Kew (UK), Universitas Negeri Papua, Manokwari (Indonesia) and the Papua New Guinea Forest Research Institute, Lae. The research was also supported by grants from the BAT Biodiversity Partnership and the Pacific Biological Foundation to the Palms of New Guinea Project.

We are particularly grateful to our partners in the DIPPDP for invaluable discussions regarding the design and contents of this book, namely Rudi Maturbongs, Jack Wanggai and Charlie Heatubun in Manokwari, and Roy Banka in Lae. Other counterparts in the Palms of New Guinea project have contributed in various ways: Ary Keim and Johanis Mogea (Herbarium Bogoriense, Indonesia), Anders Barfod and Anders Kjaer (University of Aarhus, Denmark), Scott Zona (Fairchild Tropical Botanic Garden, USA), Osia Gideon (University of Papua New Guinea, Papua New Guinea), John Dowe (James Cook University, Australia) and Ross Bayton (University of Reading, UK). Special thanks go to Ary Keim for providing the Indonesian translation, Lucy Smith and Soejatmi Dransfield (Royal Botanic Gardens, Kew) for producing the line drawings, and Anders Barfod and Scott Zona for critically reviewing the manuscript. Additional photographs were provided by Sasha Barrow, Anders Barfod, Anders Kjaer, Martin Sands, Jeff Wood and Scott Zona. We thank the staff of the many herbaria – A, AAU, BH, BM, BO, BRI, CANB, FI, K, L, LAE, MAN, MEL, NY, WRSL – who have generously given us access to their collections.

We acknowledge the Papua New Guinea Forest Research Institute, Universitas Negeri Papua, Herbarium Bogoriense, Lembaga Ilmu Pegetahuan Indonesia, the Papua New Guinea National Research Institute and the Royal Botanic Gardens, Kew for their support of our many field trips to New Guinea. Finally, we owe the success of our New Guinea expeditions to the guides, landowners, counterparts and technicians who have assisted us in exploring for palms in many exciting and remote parts of this astonishing island.

The purpose of this book

This field guide is for anyone who is interested in the palms of New Guinea, regardless of their experience or botanical expertise. Its aim is to provide easy access to information on the palms of New Guinea and act as a tool for their identification. It has been inspired by the innovative *Field Guide to the Rattans of Lao* (Evans et al. 2001). Following the style of that book, this text includes straightforward language, accessible explanations of palm morphology and terms, and carefully designed descriptive accounts that will enable the reader to identify New Guinea palms with confidence. It will be particularly useful to those people who encounter palms regularly in the field in New Guinea, such as students, foresters, field botanists, ethnobotanists, ecologists and conservationists.

The geographical coverage of this field guide includes the entire New Guinea region from the Raja Ampat Islands in the west to the Bismarck Archipelago and the Louisiade Archipelago in the east. It does not include the Aru Islands or Bougainville. Politically, it comprises the independent country of Papua New Guinea and part of the much larger Republic of Indonesia. For this reason, the book is being published in English and Indonesian language versions.

This is a guide to New Guinea palm **genera** (singular: **genus**, the classification category above the species level). The New Guinea palm flora is extremely rich and is the subject of a current research project that will result in a detailed scientific account of all species (Baker 2002a). Although this is not a guide to all the species, many of New Guinea's most common palm species are illustrated here.

How is the book arranged?

Introductory sections (pages 4–18)
Here you will find an introduction to New Guinea and its palms, as well as an introduction to palm morphology containing all the terms used in this book.

Keys (pages 19–25)
This section will help you start to identify New Guinea palms. By working through a series of criteria, you will be able to narrow down the identity of any New Guinea palm to just a single genus. If you do not have all the necessary information, you will still be able to eliminate many of the possibilities. You can then check your results against the detailed genus accounts in the next section.

Genus accounts (pages 27–103)
Each genus is described in detail with images of their habit, leaves and other features, as well as additional information such as distribution and uses. For some diverse genera, we have provided additional illustrations of a range of species.

Concluding sections (pages 104–108)
Lists of genus names, including old names that are no longer accepted, useful literature and an index are located at the back of the guide.

New Guinea and its palm flora

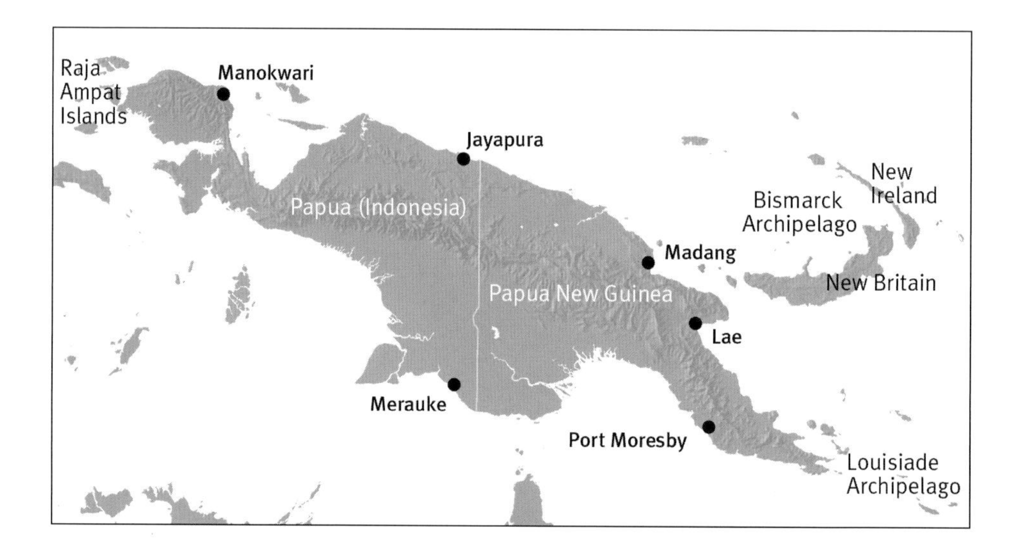

Lying just north of Australia, New Guinea is the largest tropical island in the world with a land area of over 800,000 km^2. At its northern limits, the island sits close to the equator and spans a relatively narrow range of latitudes approaching 11° S and is therefore confined within the tropics. It is a very long island of some 2500 km from west to east, extending to around 3000 km if the Raja Ampat Islands and the Louisiade Archipelago are included. In addition to large areas of lowlands, there are many mountain ranges, most of them forming part of a central spine that runs along much of the length of the island. The mountains reach great altitudes, Puncak Jaya (or Mt. Carstenz) being the highest at 5030 m with permanent ice at its summit.

Palms belong to the flowering plant family Arecaceae (also known as Palmae). They are a rather isolated and distinctive lineage within the monocotyledons, which is a large group of flowering plant families that also includes the grasses, orchids, lilies, bananas and gingers. Palms are generally very easily recognised by their distinctive fan- or feather-like leaves clustered at the tip of a straight, woody, cylindrical stem that does not branch above ground level. A small number of other plant groups resemble palms (see **Plants that look like palms**) and can cause confusion. However, palm leaves are unique and easily distinguished from all others. They develop as an undivided sheet of leaf tissue that is folded up in the main bud of the palm. When the leaf emerges from the crown, it looks like a spear, but it gradually expands and splits appear along the folds in the tissue, dividing the leaf into the complex structure with which we are familiar. You can always find evidence of the folds by looking closely at the leaf, especially at the leaflet base.

Malesia is the most important tropical region for palm diversity, containing more than 1000 of the 2300+ species in the family. New Guinea is one of three hotspots of palm diversity in Malesia, the region that ranges from Malaysia in the west to Papua New Guinea in the east. Approximately 270 palm species in 31 genera occur in New Guinea; the other two Malesian palm hotspots, the Sunda Shelf and the Philippines, have >500 and c. 140 species respectively.

New Guinea palms occur mainly in primary rainforest from sea level to 2800 m. Although many palm species occur in lowland forests, high palm diversity is also found towards the middle of the elevation range (Bachman et al. 2004). Palms thrive in a wide range of habitats in New Guinea, including mangrove, swamp forest, limestone hills and cliffs, and the flood zone of river banks.

Although many New Guinea palm species have been described scientifically, most remain poorly known. Some species have been described more than once, and therefore have more than one name, and many new species have not yet been documented by botanists. Several authors have given preliminary accounts of the New Guinea palm flora that contain much valuable information (Essig 1977, Hay 1984, Essig 1995, Ferrero 1997, Barfod et al. 2001, Baker & Dransfield 2006). However, a scientific evaluation of all palms in New Guinea has not yet been published. This is the focus of the Palms of New Guinea project, which involves collaborators from Indonesia, Papua New Guinea, UK, USA, Denmark and Australia (Baker 2002a).

The genera of palms in New Guinea reflect the diverse origins of the New Guinea flora as a whole. Many groups display links with west Malesia, such as the large genera *Calamus* and *Licuala* – New Guinea is a secondary centre of diversity for these genera, the primary centre being the Sunda Shelf region. For other genera, like *Korthalsia* and *Pinanga*, very few species are found in New Guinea, these representing the western limit of a broadly Asian distribution. Several genera, including *Clinostigma*, *Physokentia* and *Linospadix,* display strong links with Australian and Pacific palm floras. A number of genera are very species-rich in New Guinea, although small numbers of species are also found elsewhere, for example, *Calyptrocalyx*, *Cyrtostachys*, *Heterospathe*, *Hydriastele*, *Orania*, *Ptychococcus*, and *Ptychosperma*. Only three small genera are endemic to New Guinea (i.e. occur nowhere else) — *Brassiophoenix*, *Dransfieldia* and *Sommieria*. Although most palm genera in New Guinea are not restricted to the region, the majority of species are endemic.

Further information on the distribution, diversity and classification of New Guinea palms can be found in the following references: Uhl & Dransfield 1987, Dransfield & Uhl 1998, Dransfield et al. 2005. A checklist of New Guinea palm species may be downloaded from the World Checklist of Monocotyledons (http://www.rbgkew.org.uk/wcsp/).

Plants that look like palms

Cycads are gymnosperms, not flowering plants. Although the leaves and stems resemble palms, they produce cones, not flowers, and their seeds are not enclosed within fruits. The leaves do not develop like palm leaves, lacking folds and splits.

Pandans belong to the monocotyledon family Pandanaceae. They are woody trees, shrubs and climbers, but are easily distinguished by their leaves, which are strap-like, a feature which is never found in palms.

Bananas (Musaceae) are also monocotyledons. In general habit they resemble palms, but they do not have a woody stem; their stems are in fact made up of leaf sheaths. Although the leaves can be divided and appear pinnate, this is caused by wind damage. They also have distinctive inflorescences and fruit.

Bamboos are woody grasses (Poaceae or Gramineae). Although they are hard to confuse with palms, their stems resemble rattan canes and are often used for similar purposes. However, bamboo canes are usually hollow, whereas rattan canes are always solid.

Tree ferns are ferns with robust trunks that can reach several metres in height. The tall, cylindrical stem topped with large, complex leaves creates a palm-like appearance. However, tree ferns are not flowering plants, reproducing by spores rather than seeds. Their leaf development and morphology is also very different to palms.

Gingers are herbaceous plants of the forest floor. Their long, leafy shoots superficially resemble palm seedlings. A closer examination will reveal that each "leaflet" is in fact a separate leaf. The shoots do not form a rosette as a palm would, and the often colourful inflorescence and flowers are quite unlike anything found in the palm family.

An introduction to palm morphology

Habit

Palm stems are columnar. They only very rarely increase in diameter along their length. In a few palms the base of the trunk is supported by large aerial roots known as **stilt roots**.

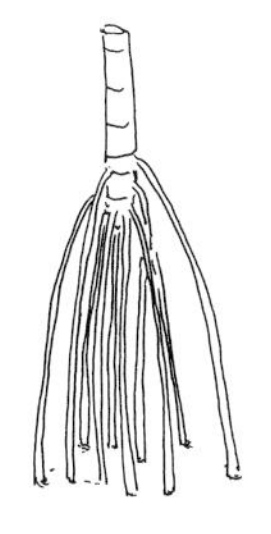

stilt roots

Above ground level, palm stems are almost always unbranched (an exception being the climber *Korthalsia* that branches in the canopy). Many palms are **single-stemmed**, while others produce shoots at the base of the trunk and are called **clustering** or **multi-stemmed**. In a few palms the stem is mostly subterranean and the palm appears to be **stemless**.

single stemmed

multi-stemmed

stemless

In *Nypa*, the mangrove palm, the stem creeps horizontally along the mangrove mud, branching by equal **forking** (**dichotomous branching**) to produce huge colonies. In New Guinea, there also many climbing palms, widely termed **rattans**.

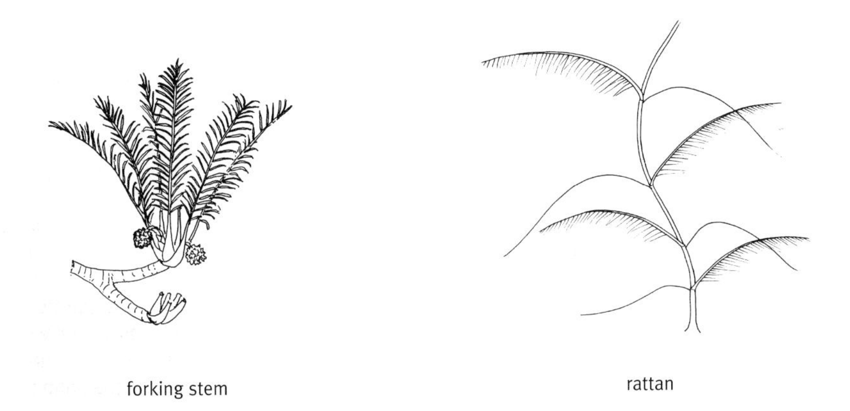

forking stem

rattan

The point at which a leaf is attached to a stem is called a **node**. The nodes can be very conspicuous on palm stems after the leaves have fallen off – they often look like rings encircling the stem. The portion of stem between two nodes is called the **internode**. The length of the internode can vary a lot between genera and species.

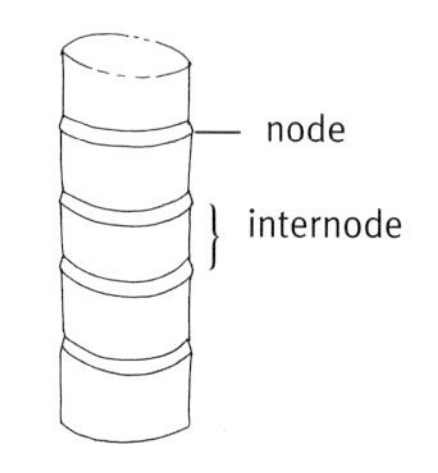

the palm stem

In many New Guinea palms, the stem tip passes into a smooth column composed of tightly enclosing tubular **leaf sheaths** (see below). This column is known as the **crownshaft** and is a very distinctive diagnostic feature of many palms (see opposite).

The leaf

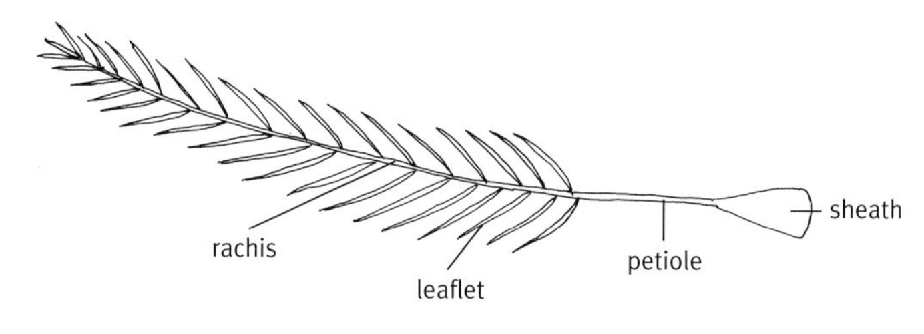

the palm leaf

The blades of palm leaves are almost always split into smaller divisions known as **leaflets** or **segments**. In New Guinea, there are three basic types of palm leaf. They can be fan-like (**palmate**) or feather-like (**pinnate**) and in one genus (*Caryota*) they are twice-pinnate (**bipinnate**), meaning that the leaflets are split again into secondary leaflets. In some pinnate and palmate leaves the blade remains undivided into separate leaflets, but is forked at the tip (**bifid**).

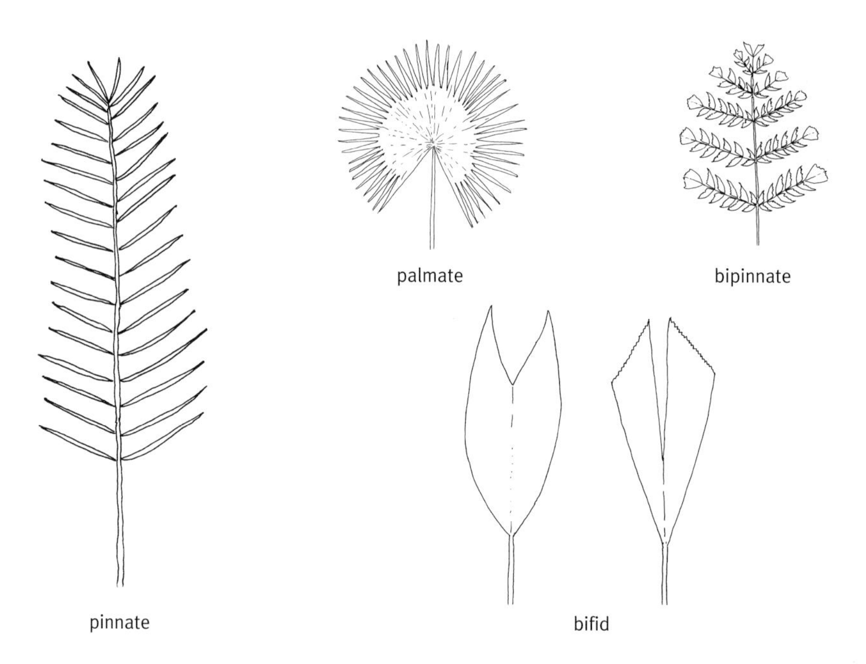

The leaf base in palms is **tubular**, forming a **sheath** around the stem. In some palms the tubular sheath **splits** or disintegrates into **fibres**. In others a deep triangular split (cleft) forms at the base of the leaf stalk (**petiole**) and sheath – this allows the stem tip to expand. In many palms the sheath remains tubular and can either be fibrous or not.

As mentioned above, strictly tubular leaf sheaths can form a **crownshaft**.

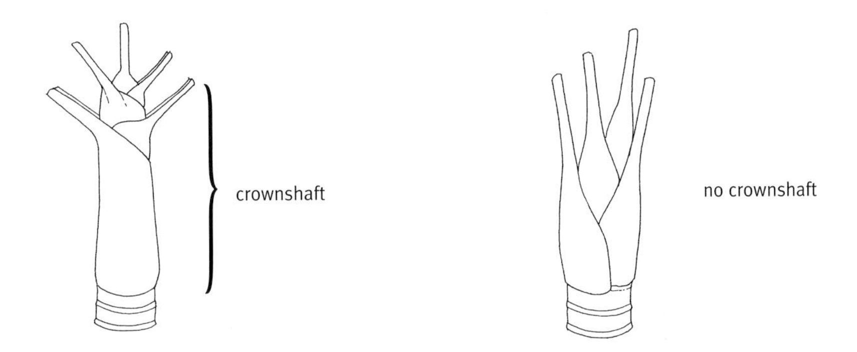

Above the sheath is the **petiole** – this can be very variable in length. Beyond the petiole lies the **leaf rachis**, the portion of the leaf that carries the leaflets. The sheath, petiole and rachis (and sometimes the leaflets themselves) are often covered with a thin or thick, felt-like layer of tiny hairs or scales (**indumentum**). In some palms (the rattans and sago palms, and many of the fan palms), the petiole is armed with **spines**. Spines may also be present on sheaths, rachis and even leaflets. Two remarkable structures are associated with the leaves in climbing palms. The **cirrus** is a long whip-like extension of the leaf rachis beyond the topmost leaflets. It is armed with hook-like spines that assist the rattan in climbing. The **flagellum** (in only some species of *Calamus*) is a spiny whip, very similar in general appearance to the cirrus, but borne on the leaf sheath rather than the leaf tip. It represents a modified inflorescence that has no flowers and functions in the same way as a cirrus. Cirri and flagella are never found in the same species (at least in New Guinea species).

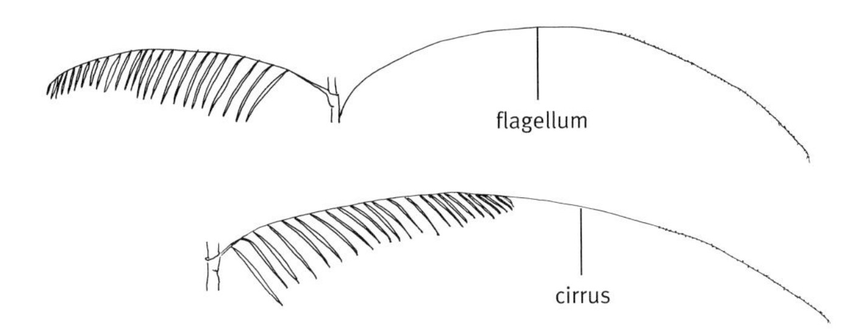

The leaf blade is divided into **leaflets** (in pinnate- and bipinnate-leaved palms) and **segments** (in palmate-leaved palms). The shape and arrangement of the leaflets and segments is often of great value in identification. As mentioned above, the young palm leaf is strongly folded within the growing tip of the palm. The young leaves in the growing tip (**shoot apex**) are often, but not always, edible, and are known as **palm cabbage** or **palm heart**. As the leaf expands, it splits to form the segments or leaflets. If the splits occur on the lower folds, then the resulting segments or leaflets are Λ-shaped in cross section at the base and are termed **reduplicate.** If the blade splits on the upper folds, then the resulting leaflets and segments are V-shaped in section and termed **induplicate.** If the splits do not occur on every lower or upper fold then the resulting segments or leaflets consist of more than one fold (**multifold**).

induplicate reduplicate

Leaflets may be arranged **regularly** along each side of the leaf rachis, or **irregularly**, often in groups.

regular irregular

The orientation of the leaflets is also diagnostically useful; they may hang down on either side of the rachis (**pendulous**), they may be held **horizontally** or may point upwards (**ascending**).

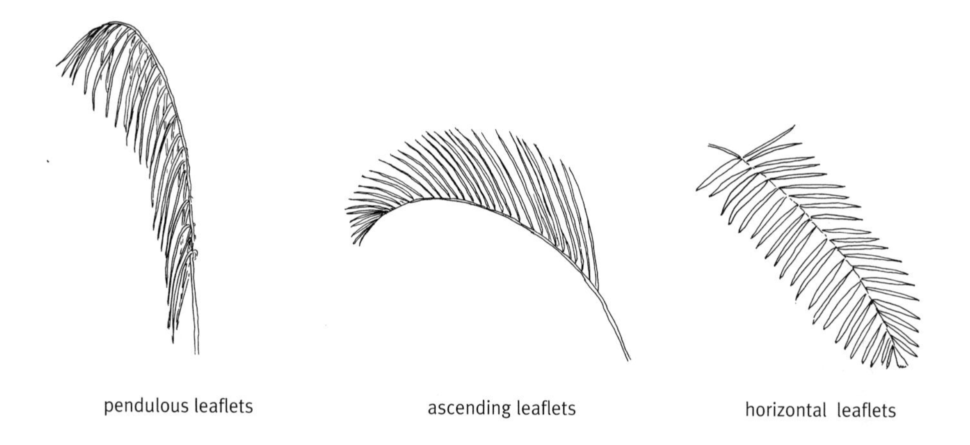
pendulous leaflets ascending leaflets horizontal leaflets

Leaflet and segment shape varies from narrow with parallel sides (**linear**) to wedge-shaped. The leaflets tip can be **pointed, lobed, toothed** (with small regular teeth) or **jagged** (appear as if torn or bitten off). In some palms the leaflet margins and/or the main veins, either on the upper or lower surfaces, or on both surfaces, may carry small **spines** or **bristles**.

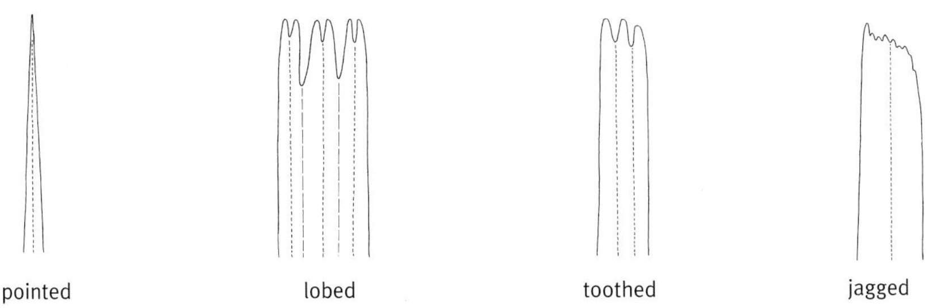

The inflorescence

Palm inflorescences emerge below, between or above the leaves. Palms with crownshafts usually produce their inflorescences **below the leaves**. Palms without crownshafts usually produce their inflorescences **between the leaves**. In a few New Guinea species, many inflorescences are produced at the same time in the axils of reduced leaves at the very tip of the stem. In such species the resulting mass of inflorescences appears to be a single large inflorescence borne **above the leaves**. In these species the individual stem dies after flowering and fruiting. If the palm is single-stemmed, then the whole palm dies; if the palm is multi-stemmed, then shoots at the base of the trunk replace the dying stem.

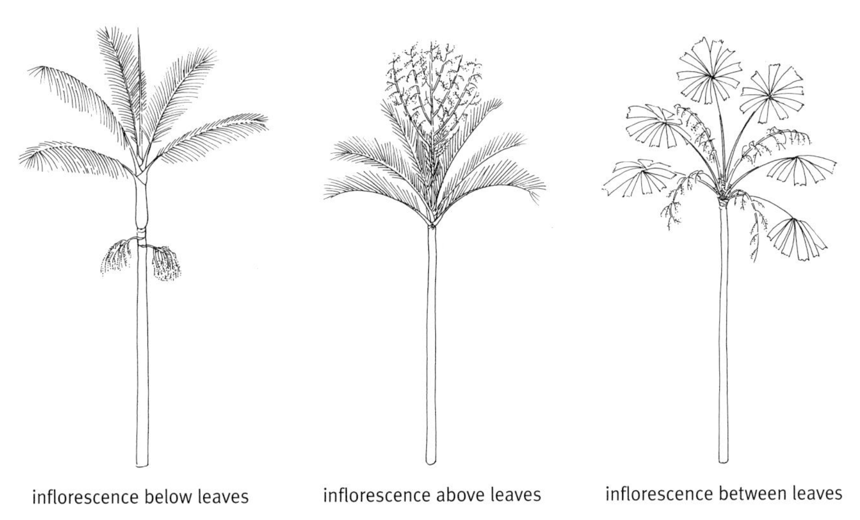

The inflorescence starts its life as a bud, enclosed in one or more tubular sheaths or **bracts**. The first bract on the inflorescence is referred to as the **prophyll**. It may be flattened or rounded in section, and often has a long point or **beak**. The basal unbranched portion of the inflorescence (the inflorescence stalk) as far as the first branch is called the **peduncle**. Besides the prophyll, the peduncle may carry one to several additional bracts, referred to as **peduncular bracts**. The peduncular bracts and the prophyll may fall off the inflorescence as it expands or they may persist. The upper branched portion of the main inflorescence axis is called the **rachis**. The rachis may have inconspicuous bracts or they may be very similar in form to the peduncular bract(s). Collectively, the prophyll, peduncular and rachis bracts are sometimes termed the **primary bracts**. Inflorescence bracts and branches may be covered in various types of hairs or scales (**indumentum**). The bracts may also be spiny.

Palm inflorescences are usually highly branched, although they may be unbranched (**spicate**). We describe the degree of branching in terms of numbers of **branching orders**.

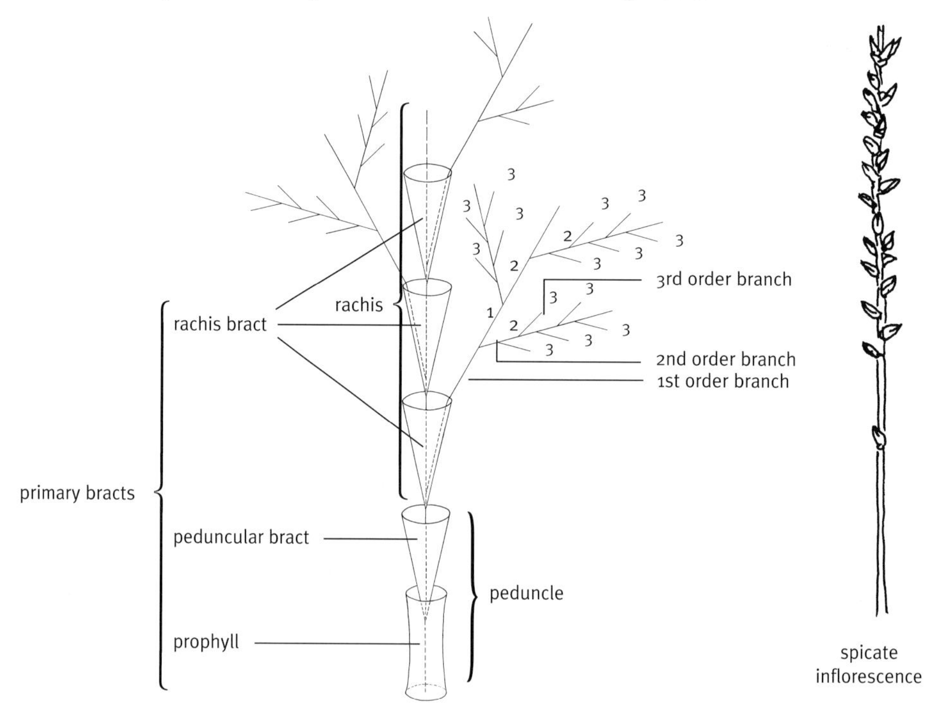

The peduncle and rachis represent order zero. Any branches from the rachis are counted as first order branches and if no further branches are present, the inflorescence is said to branch to one order. If more branches arise from the first order branches, these are called second order branches and the inflorescence is said to branch to two orders, and so on. Each additional order is counted sequentially thereafter. Some New Guinea genera produce inflorescences with up to five orders of branching.

The branches of the inflorescence that carry the flowers are referred to as **rachillae**. How the rachillae are displayed is often useful for identification, for example they may be **pendulous** or **spreading**. In a spicate inflorescence, the flowers are borne directly on the main axis of the inflorescence in its upper part.

Flowers

Flowers may be borne on the surface of the rachillae or they may be set into depressions or **pits** in the rachillae. The flowers emerge from the pits as they mature.

Palm flowers may be **hermaphrodite** (both fertile male and female parts are present). More usually, the flowers are of a single sex or **unisexual**. Where both male and female flowers are borne on the same palm we refer to this condition as **monoecious**. Where unisexual flowers are borne on different trees, we end up with separate male and separate female trees. Such palms are referred to as being **dioecious**.

How the individual flowers are arranged along the rachilla is of importance in the classification of palms. In some, they are single (or **solitary**) at each point of attachment. In others they are arranged in **pairs**. In many palms they are arranged in threes (**triads**) consisting of a central female and two lateral male flowers; while in a few palms, the flowers are arranged in tight **clusters** of 2–7 flowers. In those palms that have triads, the male flowers can open before the females, or the females open before the males. This separation of opening times helps to ensure cross-pollination.

Fruit

The fruit develops after fertilization of the female flower. Usually the fruit is more or less symmetrical. Occasionally (in *Orania*), when more than one seed develops, the fruit can have two or three lobes, each with an enclosed seed (**bilobed, trilobed**). The **stigmatic remains** (the female receptive surface) can be at the tip of the fruit (**apical**), to one side of the tip (**subapical**), at the side (**lateral**) or the base (**basal**). Usually the surface of the fruit is **smooth.** In the rattans and sago palms, the outer surface of the fruit is covered in neat vertical rows of **overlapping scales** that give the fruits a most distinctive appearance. Very rarely (in *Sommieria*) the fruit surface is broken into **corky warts.**

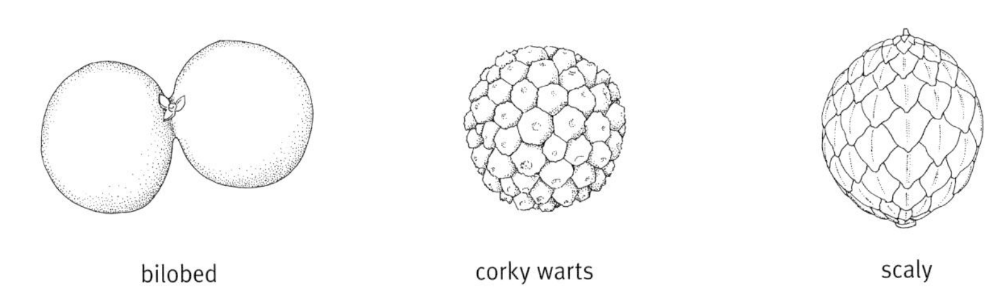

bilobed | corky warts | scaly

Just inside the outer covering of the fruit lies the **fruit flesh,** which can be **thin** or **thick, juicy,** or **dry** and **fibrous** (when it is sometimes called **husk**). In some genera (e.g. *Arenga, Caryota, Drymophloeus*) the flesh contains microscopic **needle crystals** that make the flesh and juice extremely irritating to the skin. Within the flesh lies the **endocarp,** a usually hard layer that surrounds the seed. The endocarp can be **smooth,** covered in **fibres, ridged** or **grooved.** In the coconut there are three distinct **eyes** in the endocarp through one of which the young seedling will eventually emerge on germination. In *Orania* the endocarp also has a swelling at its base that we refer to as a **basal heart-shaped button**; its function is unknown.

Inside the endocarp lies the **seed.** In some rattans the outermost layer of the seed coat is thick and fleshy (**sarcotesta**) functioning in the same way as the juicy fruit flesh of other palms. The young growing point of the seed (**embryo**) is within the hard body of the seed that is called **endosperm.** The endosperm is very hard, usually white in colour and is either uniform in texture (**homogeneous**) or penetrated by thin brown lines of seed coat tissue (**ruminate**).

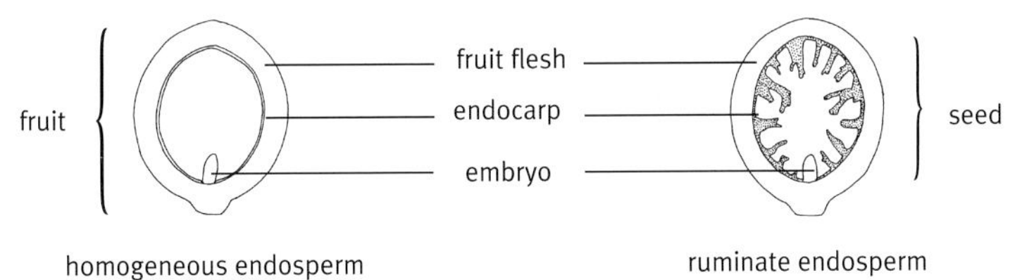

homogeneous endosperm | ruminate endosperm

When the seed germinates, the first leaf (**eophyll**) that is produced often has a characteristic shape.

Shape terms used in this book

Throughout this book we use special terms to describe the shapes of some palm parts.

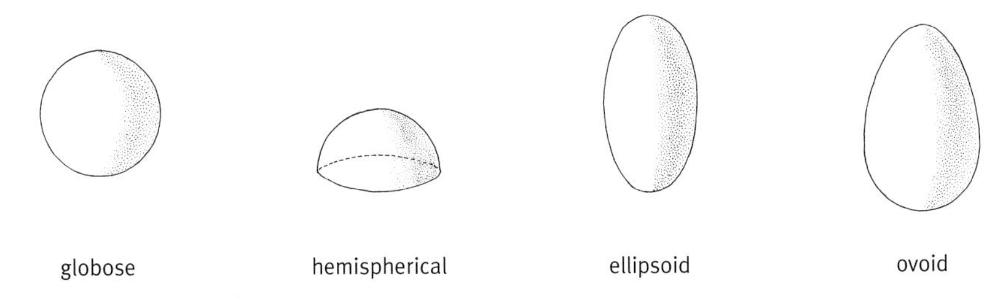

globose | hemispherical | ellipsoid | ovoid

Further reading

If you want to know more about palm morphology and terminology, you will find excellent discussions and glossaries in the following references: Uhl & Dransfield 1987, Tomlinson 1990, Dransfield & Beentje 1996, Jones 1995.

Palm specimens and how to collect them

The **identification**, scientific classification (**taxonomy**) and naming (**nomenclature**) of plants is dependent on reference collections of preserved specimens. For centuries, botanists and explorers have collected specimens during expeditions all over the world. These usually consist of portions of plants that are pressed in newspaper and dried over a gentle heat source. Plant specimens are then deposited at a **herbarium**, where they are usually stuck (or **mounted**) on a piece of stiff paper with field notes and identifications and stored in a systematic order in cupboards. The specimens may be used for several purposes:

1. Identification – you can compare new collections against expertly named specimens that are already in the herbarium.
2. Taxonomy – by studying the specimens in detail, you can improve our understanding of how genera and species are defined.
3. Nomenclature – the description of a new species requires that a single herbarium specimen – the **type** (usually a dried pressed specimen, but can also include pickled material preserved in spirit) – is cited as a reference point for the name of the species.

Our understanding of the limits of species (and indeed genera) of New Guinea palms is thus very heavily dependent on the many specimens found in herbaria scattered around the world. The more specimens of a given species that are available for study, the better the understanding of the range of variation of that species.

Unfortunately, palms are not straightforward plants to collect and represent in the herbarium. They are often far too large to fit on herbarium sheets and are often bulky and spiny. Consequently, many palm specimens are really inadequate in representing the form of the living palm, and often historic type specimens, which are so important for naming palms, are difficult to interpret. However, given patience, plenty of time and common sense it is possible to make good specimens in the field that are of real value. We provide some guidelines here on how best to go about making a specimen in the field (a more detailed account may be found in Dransfield 1986). As the palm flora of New Guinea is still poorly known, the chances that you may encounter undescribed (new) species is actually quite high, and your efforts in making a good collection may well be rewarded by new discoveries.

Collecting palms

Before you attempt to make a collection, **you must ensure that you have permission to do so from local inhabitants, landowners and relevant government authorities**.

Make sure you have plenty of time to prepare a good specimen. While some smaller palms are quick and easy to collect, tackling a large tree palm may take two hours or more. You will need the following equipment:

- Secateurs
- Notebook
- Pencil (never use ink – it may dissolve in water or alcohol)
- Jeweller's tags (hanging labels)
- Bush knife (machete or parang)
- Leather gardening or industrial gloves
- Tape measure
- Newspaper

In addition, a folding pruning saw can be immensely useful, particularly if you are not very skilled with a bush knife. A camera to take photographs of the living palm is of immense value. A GPS (global positioning system) unit is a very important tool that is increasingly available – it will enable you to pinpoint the exact locality of your collection by satellite.

The main aim of your collection should be to represent the living palm in such a way that someone else can reconstruct in their mind what the palm looked like from your specimen and accompanying notes when it is deposited in the herbarium. In general, **there is little point in collecting a palm that has neither flowers nor fruit** yet sometimes such sterile collections are useful in recording the presence of a species. However, **rattans are the exception to this rule** – they have many informative features in the leaf and leaf sheath and sterile collections are still very worthwhile.

Field notes

You should start by writing field notes. The following points are essential for your notes, but you should feel free to describe more details. **Remember, you are aiming to document details that will not be obvious in the preserved herbarium specimen itself.**

1. Write your name (and any other collectors helping you) and a unique number. Every part of the specimen should be labelled with this information – it will become the unique reference for the collection. For example, the number *Baker 1066* is given to material of *Dransfieldia micrantha* collected by William Baker and his collaborators on 26 February 2000 in Papua, Indonesia. It consists primarily of herbarium material, but additional material collected at the same time (fruits and flowers pickled in alcohol, living seeds for cultivation, leaf material for DNA extraction and photographs) was given the same number to link it all together with the collection data. If you have never collected plants before, start to use a simple numerical sequence for your specimens (e.g. 1, 2, 3, 4, and so on), but never use the same number for more than one collection.
2. Note the date.
3. Note where you are, in what sort of forest/habitat, and at what altitude.
4. Note what the palm is called in the language used in that area, and if it has any uses.
5. Is the stem single or clustered?
6. How tall are the stems and what is their diameter?
7. How long are the internodes on the stem?
8. Are any special roots visible?
9. How many leaves are there in the crown and how long are they?
10. How long and wide is the sheath, and do they form a crownshaft?
11. How long and wide is the petiole and leaf rachis?
12. Count the number of leaflets on each side of the leaf rachis and say whether they are arranged regularly or irregularly or grouped in any special way, and whether they are held in the same or several planes.
13. Measure an inflorescence and describe where it emerges – above, between or below the leaves.
14. Measure the length of the peduncle and the rachis and count the number of orders of branching.
15. Describe the colour of the bracts, branches and flowers or fruit.

All these notes already give much information on the living plant, but you can describe anything else that you think might be of interest. It is better to collect too much information than too little. Your notes will be really valuable documentation for the specimen, especially if supplemented by a photograph.

Making a specimen

As you collect the palm, you should be thinking all the time about cutting pieces of the palm to a size conforming approximately to that of a folded newspaper. Remember, if you make specimens that are too big for herbarium sheets (c. 30 × 40 cm) it will be difficult to store your material. The following instructions will help you prepare a complete and informative specimen:

1. **Stem** – If the stem is slender then you may take a sample of stem. If very robust, it is usually possible to cut out a thin strip of stem that includes the outer surface.

2. **Sheath** – Take a whole leaf sheath and split it down the middle. If the sheath is very large, you may cut the halves into fragments representing the base and the tip of the sheath.
3. **Leaf** – Remove a whole leaf. If the leaf is small, you may be able to collect it whole, folding it to fit the sheet. In most cases, however, you will need to cut it up, keeping only specific parts, which may also require folding to fit a herbarium sheet. For pinnate leaves, take a portion of the base of the petiole, a portion of the tip petiole with the first leaflets, a middle section with rachis and leaflets, and the tip of the leaf. You may remove the leaflets from one side if they are too big. For fan leaves, take a portion of the base of the petiole, a portion of the tip petiole with lower segments on one side, then remove a portion from the centre of the leaf and from the side.
4. **Inflorescence** – If the inflorescence is small and can be folded conveniently to fit the size of the folded newspaper, then there is no need to cut it further. If it is large, then cut the inflorescence to provide a basal portion, a mid section and a tip.
5. **Flowers and fruit** – Make sure you have good flowers and/or fruit in the specimen. You may not be able to obtain both from a single palm. If flowers are not visible, you may be able to find some inside an inflorescence bud. If no ripe fruits are available, you may be able to find fruits or seeds on the ground around the palm – even germinating seedlings contain useful information.
6. **Rattans** – Climbing palms can be even more challenging to collect because of their spines, but the general principle is the same. In the case of rattans, however, it is important to collect the climbing whips and to note whether they arise from the leaf tip (cirrus) or leaf sheath (flagellum). Do not attempt to remove the sheath from the stem. Simply cut a length of stem with sheaths including at least one leaf base.
7. **Label your material** – Because palm collections contain so many parts, they must be labelled properly to avoid mix-ups. Once everything is cut up you are ready to label the specimen. Each piece of the collection should be labelled with a tag on which your name and unique collection number is written.
8. **Press the specimens** – Once labelled, the material can be pressed between newspaper. If you are on a day trip, then it will be easy to take your specimen back to base and then dry it. If you are on an extended trip, then the specimen can be preserved for a few weeks by placing bundles of specimens in robust plastic bags and soaking them with about 0.5 l of methylated spirit or 70% ethanol.
9. **Spirit and DNA collections** – Really comprehensive modern collections also include pickled material of delicate parts (such as flowers or fruit) – these may be preserved in methylated spirit or 70% ethanol. In addition, a small amount (c. 5 cm^2) of young leaf material dried rapidly in silica gel desiccant can be used for DNA extraction.
10. **Make duplicates** – We have described here how to make a single specimen. It is usually possible to make more than one duplicate specimen from the same individual palm. This significantly increases the value of your collection effort because you can send duplicate specimens to other herbaria, so that specialists elsewhere can have easy access to your material and provide you with accurate identifications. It also provides insurance against accidental loss of specimens.

Where to see New Guinea palms

Living palms

Many New Guinea palms are spectacular ornamentals and because of this they are popular with growers of palms and are sometimes well represented in botanic garden collections. The most important historic living collections of New Guinea palms were to be found in Kebun Raya Bogor in Indonesia. In the mid 19th century, collectors introduced many species into the Bogor gardens and those species that were undescribed or little known were featured in an important series of articles by Scheffer and Beccari in *Annales du Jardin Botanique de Buitenzorg* (the old name for Bogor). A few of these old palms may still survive in Bogor, but what is more certain is that generation after generation of seed from these original collections were planted in the garden and distributed elsewhere. In recent years, collecting by the botanic garden's staff and palm botanists from the

Herbarium Bogoriense has increased the representation of New Guinea palms in the garden and Kebun Raya remains a marvellous place to get to know the genera of New Guinea palms.

In New Guinea itself, the most significant collection of New Guinea palms is in the Botanic Garden at the Forest Research Institute in Lae, but other collections, for example in Manokwari, are being developed. Outside the SE Asian region, major collections of New Guinea palms can be seen in Flecker Botanic Gardens in Cairns and the Townsville Palmetum, both in Australia, and several of the botanic gardens in Hawaii (especially Hoomaluhia and the Lyon arboretum). In the continental USA, Fairchild Tropical Botanical Garden and the Montgomery Botanical Center, both in Miami, Florida, have splendid palm collections that contain many New Guinea species. In Europe, the most significant living collection of palms is in the Palm House at the Royal Botanic Gardens, Kew, in London.

Palms in herbaria

Although living collections may help you to know and understand palms better, they are not the best resource for identification. For this purpose, you need to use a reference collection of herbarium specimens collected from the wild. Within New Guinea, important palm herbarium collections are available in the Forest Research Institute herbarium in Lae, the University of Papua New Guinea in Port Moresby and in the Biodiversity Study Centre, Universitas Negeri Papua, Manokwari. Elsewhere in Indonesia, the Herbarium Bogoriense, Bogor, houses a significant collection with many historical specimens. Outside the region, the most important herbaria for New Guinea palms are the Australian National Herbarium, Canberra; the Queensland Herbarium, Brisbane; the Botanical Institute in Florence; the National Herbarium of the Netherlands (Leiden branch); the Royal Botanic Gardens, Kew; the Harvard University Herbaria and the L.H. Bailey Hortorium, Cornell University. Several other herbaria house a growing number of recently collected New Guinea palm specimens, for example the University of Aarhus, Denmark and Fairchild Tropical Botanic Garden in Florida.

Identification Keys

Keys are widely used tools in plant identification. By comparing a plant against alternative sets of features in the key, it is possible to narrow down its identity step by step, until only one possibility remains. Keys rarely work perfectly, most often because the specimen in hand may lack the information required by the key; for example, the key may focus on features of the flowers, but your specimen may be in fruit only. In our keys, we have included the features that you are most likely to observe or that will most often be available to you. There is no substitute for careful observation. Familiarise yourself with the criteria that are used most often and with the basic features of the palm you want to identify. Try to answer all these questions before you start working through the key:

- What is the habit of the palm and the diameter of its stem?
- What is the overall shape of the leaf?
- What do the leaflet tips look like, both in the middle of the leaf and at the apex?
- Does the palm have a crownshaft?
- What is the overall shape of the inflorescence?

If a palm is too tall and you cannot see the crown properly, look on the ground at the base of the trunk – you may find fallen leaves, inflorescences, flowers or fruit that will help you.

How the keys work

After making general observations, refer to the General Key – this key allows you to decide which group your palm belongs to and will direct you to another key for identifying a specific group of genera. Compare the palm against the list of alternatives on the left hand side of the table. In this case, you must decide if “Leaves palmate”, “Leaves bipinnate” or “Leaves pinnate” best matches the palm that you are identifying. Having decided, move to the next column of criteria to the right. If your palm matched "Leaves pinnate", you have eliminated the other options and therefore the other rows in the key. You now reach a new set of choices – "Climbing palms" or "Short to tall palms, not climbing". Address those features and, again, move on to the right. Eventually, you will reach a final answer that will direct you to the next key.

You may reach your answer more quickly for some palms than for others. For example, if you decided that the palm matched "Leaves palmate", you will immediately arrive at an answer – you are identifying a fan palm and you should proceed to Key A. In some cases, you will find that a palm genus appears more than once in a key. This is because of the wide range of variation in some genera.

Example:

- We have found a 20 m tall tree palm with pinnate leaves, so we eliminate "Leaves palmate" and "Leaves bipinnate" from the General Key. We are asked next about the palm's habit. It is not a climbing palm, but matches "Short to tall palms, not climbing". We move on to the next column to the right.
- The key now asks if the palm has a crownshaft or not. If you do not fully understand this or any other term, check the **introduction to palm morphology** as you work through the key. Our palm has no crownshaft and so the key advises us to move on to Key D.
- Key D begins by asking about the appearance of the leaflet tips. Our palm has pointed leaflets, so corresponds best with the criterion "Leaflet tips pointed, lobed or toothed, but not jagged".
- Next we must check for the presence of spines. Our palm has fine spines on the leaf rachis and sheath and therefore matches. We also found some scaly fruit on the ground under the palm. Our observations agree with the criterion "Leaflets, rachis, petiole or sheath spiny, fruit scaly". We are now very close to reaching an answer.
- The fruit of this palm are small. It was also single-stemmed and we could see the inflorescences among the leaves. The palm matches very well the criterion "Single-stemmed, stems not dying after flowering, inflorescences between or below leaves, fruit ≤1 cm long" – we have successfully identified it as *Pigafetta*.
- To make certain of the identification, we check the detailed genus treatment, comparing our palm against the descriptions and illustrations. If we are concerned that we have made a mistake, we can check the "Confused with" section to find out what other options there may be. If you have found a palm outside the distribution range given in the treatment, you should check again that you have identified it correctly, but do not assume that you have made a mistake. You may have discovered a genuine new distribution record.

General Key

Leaves palmate			**Fan palms** See Key A
Leaves bipinnate			**Bipinnate palms** ***Caryota* only**
Leaves pinnate	Climbing palms		**Rattans** See Key B
	Short to tall palms, not climbing	Crownshaft present	**Pinnate palms (with crownshafts)** See Key C
		Crownshaft absent	**Pinnate palms (without crownshafts)** See Key D

Key A – Fan Palms

Stems ≤11 cm diameter, leaf blade usually split to base forming wedge-shaped segments			***Licuala*** p. 32
Stems ≥12 cm diameter, leaf blade not split to base and lacking wedge-shaped segments	Petiole base lacking large triangular cleft		***Livistona*** p. 36
	Petiole base with large triangular cleft	Petiole not spiny, inflorescences between leaves, rachillae fat >2 cm diameter, fruit ≥12 cm long, stems not dying after flowering	***Borassus*** p. 28
		Petiole spiny, inflorescences above leaves, rachillae thin <0.5 cm diameter, fruit ≤2 cm long, stems dying after flowering	***Corypha*** p. 30

Key B – Rattans

Leaflets diamond-shaped, leaflet tips jagged, stems dying after flowering		***Korthalsia*** p. 46
Leaflet not diamond-shaped, leaflet tips pointed, stems not dying after flowering	Climbing whips on leaf tips or leaf sheaths, primary bracts not splitting to base and never dropping off, widespread	***Calamus*** p. 40
	Climbing whips on leaf tips only, primary bracts splitting to base and usually dropping off, far west New Guinea only	***Daemonorops*** p. 44

Key C – Pinnate palms with crownshafts

<table>
<tr><td rowspan="5">Leaflet tips jagged</td><td colspan="3">Inflorescence branches swept forward (rarely unbranched), resembling a brush or horse's tail</td><td>Hydriastele (in part) p. 76</td></tr>
<tr><td rowspan="4">Inflorescence branches widely spreading</td><td rowspan="2">Fruit >3 cm long, endocarp conspicuous grooves and ridges</td><td>Stem diameter ≤7 cm, leaflets wedge-shaped, ripe fruit orange, endocarp pale</td><td>Brassiophoenix p. 56</td></tr>
<tr><td>Stem diameter ≥9 cm, leaflets linear, ripe fruit red, endocarp black</td><td>Ptychococcus p. 96</td></tr>
<tr><td rowspan="2">Fruit <3 cm long, endocarp smooth or grooved</td><td>Stilt roots absent, prophyll dropping off as inflorescence expands</td><td>Ptychosperma p. 98</td></tr>
<tr><td>Stilt roots usually present, prophyll not dropping off as inflorescence expands</td><td>Drymophloeus p. 70</td></tr>
<tr><td rowspan="5">Leaflet tips pointed, lobed or toothed, but not jagged</td><td rowspan="5">Stem >12 cm diameter</td><td rowspan="2">Inflorescence branches swept forward (rarely unbranched), resembling a brush or horse's tail</td><td>Fruit symmetrical, stigma at apex</td><td>Hydriastele (in part, probably H. costata) p. 76</td></tr>
<tr><td>Fruit asymmetrical, stigma to one side of apex</td><td>Clinostigma p. 62</td></tr>
<tr><td rowspan="3">Inflorescence branches widely spreading</td><td>Leaf not strongly arching, flowers in pits, fruit ≤1.2 cm long, black, endosperm homogeneous</td><td>Cyrtostachys (in part) p. 66</td></tr>
<tr><td>Leaf not strongly arching, flowers not in pits, fruit 1.5–3.5 cm long, yellow or red, endosperm ruminate</td><td>Rhopaloblaste (in part) p. 100</td></tr>
<tr><td>Leaf strongly arching, flowers not in pits, fruit >4 cm long, red, endosperm ruminate</td><td>Actinorhytis p. 48</td></tr>
</table>

continued opposite

Key C – continued

Leaflet tips pointed, lobed or toothed, but not jagged (continued)	Stem ≤12cm diameter	Some leaflets with lobed tips, mainly at leaf apex	Female flowers and fruits at base of inflorescence branches only, fruiting inflorescence often club-like	***Areca*** p. 50
			Female flowers and fruits from base to tip of inflorescence branches, fruiting inflorescence not club-like	***Pinanga*** p. 94
		All leaflets pointed	Single- or multi-stemmed, stilt roots absent, flowers in pits, fruit black, endosperm homogeneous, widespread	***Cyrtostachys*** (in part) p. 66
			Single-stemmed, stilt roots absent, flowers not in pits, fruit yellow or red, endosperm ruminate, widespread	***Rhopaloblaste*** (in part) p. 100
			Usually multi-stemmed, stilt roots absent, flowers not in pits, fruit black, endosperm ruminate, west New Guinea	***Dransfieldia*** p. 68
			Single-stemmed, stilt roots present, flowers not in pits, fruit black, endosperm ruminate, New Britain	***Physokentia*** p. 90

Key D – Pinnate palms without crownshafts

Leaflet tips jagged	Multi-stemmed, stems dying after flowering, leaflets V-shaped in section		***Arenga*** p. 54
	Single-stemmed, stems not dying after flowering, leaflets Λ-shaped in section		***Orania*** p. 86
Leaflet tips pointed, lobed or toothed, but not jagged	Leaflets, rachis, petiole or sheath spiny, fruit scaly	Single- or multi-stemmed, stems dying after flowering, inflorescences above leaves, fruit ≥2 cm long	***Metroxylon*** p. 82
		Single-stemmed, stems not dying after flowering, inflorescences between or below the leaves, fruit ≤1 cm long	***Pigafetta*** p. 92
	Spines absent, fruit smooth or with corky warts	Stem branching by forking, horizontal, abundant in mangrove forest	***Nypa*** p. 84
		Single-stemmed, stem erect, to 30 cm diameter, coastal, cultivated for coconuts	***Cocos*** p. 64
		Inflorescences unbranched, peduncular bract at tip of peduncle, fruit smooth	***Calyptrocalyx*** p. 58

continued opposite

Key D – continued

Leaflet tips pointed, lobed or toothed, but not jagged	Spines absent, fruit smooth or with corky warts (continued)	Single- or multi-stemmed, stems erect (rarely stemless), to 10 cm diameter	Inflorescences unbranched, peduncular bract near base of peduncle, fruit smooth	***Linospadix*** p. 80
			Inflorescences branched, peduncular bract near base or tip of peduncle, fruit smooth	***Heterospathe*** p. 72
			Inflorescences branched, peduncular bract at tip of peduncle, fruit with corky warts	***Sommieria*** p. 102

The genus accounts

Each of the 31 genera of palms in New Guinea is described in detail in a double-page account. For seven of these genera (*Areca*, *Calamus*, *Calyptrocalyx*, *Heterospathe*, *Hydriastele*, *Licuala* and *Orania*), an additional double-page account illustrating a range of species is given after the main genus account to highlight the range of variation in these large and diverse groups. The genus accounts are divided into groups according to the general key: fan palms, bipinnate palms, rattans and pinnate palms. Within each group, we have arranged the genera in alphabetical order. Don't forget – **the introduction to plant morphology contains simple explanations of all the technical terms that we have used**, some of which are illustrated.

At the top of each account, the general key group and the genus name are given. Below that, you will find a list of five of the most important characters for identification - these are used consistently across all genera making comparisons easy. A box entitled "Look for:" follows, in which a unique combination of three main features of the genus is given in simple language. The guide is designed so that you can identify the genus using this information alone, although you will almost certainly want to check the detailed description and illustrations.

A map shows you the distribution of the genus in New Guinea alongside details of the global distribution, habitat, number of species, taxonomic accounts and a summary of uses. Because of the enormous diversity of languages in New Guinea, we have not provided local names. To give an accurate listing of local names would require an immense amount of new work in the field and herbarium. In addition, you should **never base an identification on a local name** because the use of names, though often consistent, can vary from place to place, or person to person.

Next, a description details precise characteristics and dimensions of the habit, leaf, inflorescence, flowers and fruit. The language is more technical in this section. Remember – **the description is focused on New Guinea species of each genus only**. It does not include the features of any species that occur outside the geographical scope of this book. The description is illustrated with images that will help you to understand the description. In particular, we provide line drawings of examples of leaflet morphology – this feature is especially useful for many genera.

The account concludes with a section entitled "Confused with:". You will find a list of other palms here which may be confused with the genus in question. We have listed genus names, followed by the key characters with which they can be distinguished from the genus described in the account.

If your palm does not match the morphology or distribution described in the account, check the identification against other genera, but do not assume that you are wrong. You may have found a new record or a new morphological aspect of a known species, perhaps even a new species. This book is based on detailed observations by the authors in the herbarium and in the field, but we cannot include variation that we have not yet observed ourselves. We will be pleased to hear about your new discoveries.

And lastly - **GOOD LUCK!**

opposite: *Sommieria leucophylla* fruit, Timika, Papua
Photo: W. J. Baker

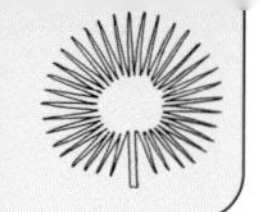

Borassus

Tall – palmate – petiole base cleft – no spines – leaf segments not wedge-shaped

Look for:

- Massive, single-stemmed fan palm with distinctive triangular cleft at base of petiole.
- Petioles very sharp, but not spiny.
- Inflorescences with fat branches, fruit very large (12–15 cm long).

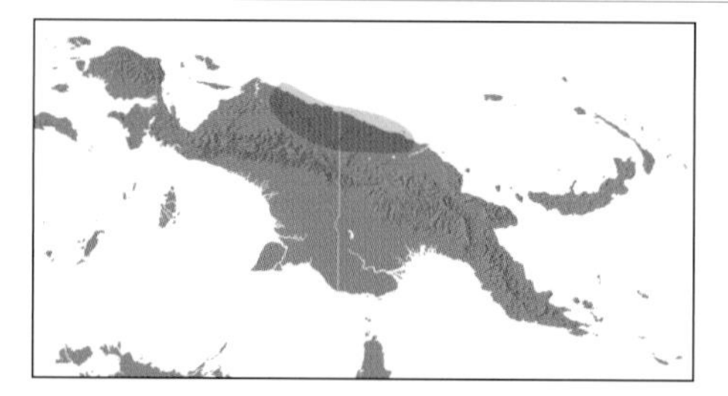

Distribution
Africa, Madagascar, India to New Guinea, in New Guinea known only from the north.

Habitat
Lowland rainforest, sea level to 100 m.

Number of species
6, of which 1 in New Guinea.

Taxonomic accounts
Bayton 2005.

Uses
Stem for construction; leaves for thatch.

Habit
Robust, single-stemmed fan palm, height to 25 m, stem diameter c. 35 cm, dioecious.

Photo: W. J. Baker

Borassus heineanus, Lae Botanic Garden, PNG

Photo: W. J. Baker

Borassus heineanus leaf, Lae Botanic Garden, PNG

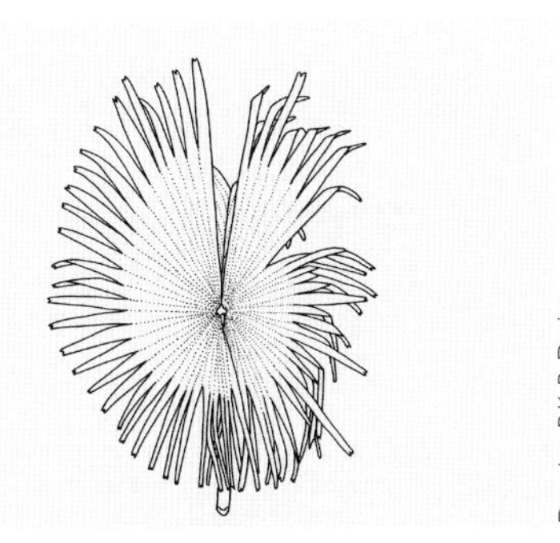

Drawing: P. K. R. Davies

Borassus leaf

Photo: W. J. Baker

Borassus heineanus male inflorescence, Lae Botanic Garden, PNG

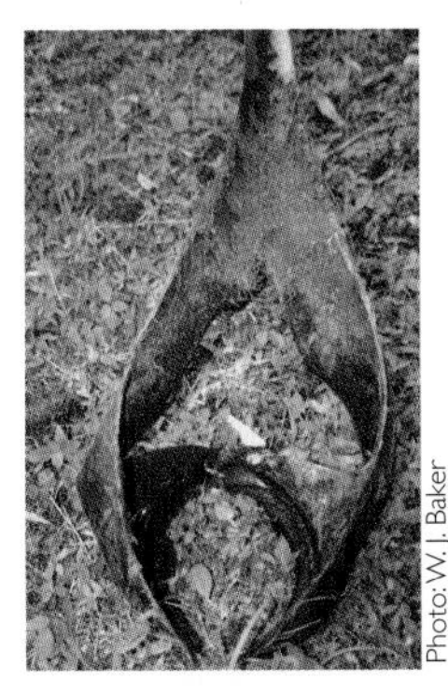

Photo: W. J. Baker

Borassus heineanus leaf sheath showing distinctive triangular cleft, Lae Botanic Garden, PNG

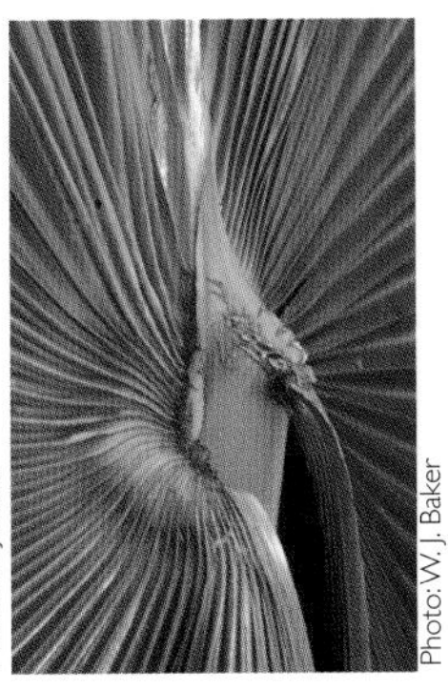

Photo: W. J. Baker

Borassus heineanus leaf detail, Lae Botanic Garden, PNG

Leaf palmate, 450 cm long, 20–30 in crown.

Sheath splitting ± to the base opposite the petiole.

Petiole robust, 150-300 cm long, with a conspicuous triangular cleft at the base, the margins very sharp, but not spiny.

Segments 80–90, with pointed tips, the blade split to c. one-third or more of its radius, blade to 180 cm long.

Inflorescence between the leaves, the male inflorescence branched to 2 orders, the female inflorescence unbranched or with a single branch near the base to 130 cm long, pendulous.

Prophyll and peduncular bract similar, not dropping off as inflorescence expands.

Peduncle longer than inflorescence rachis, to 80 cm.

Rachillae very robust.

Flowers male flowers in groups developing in pits hidden by overlapping bracts on a thick sausage-like rachilla, the flowers emerging one at a time, female flowers borne singly, very much larger, surrounded by overlapping rounded bracts.

Fruit black, 12–15 cm × 8–10 cm, stigmatic remains apical, flesh yellow, fibrous.

Seeds 1–3, each inside its own endocarp, endosperm homogeneous.

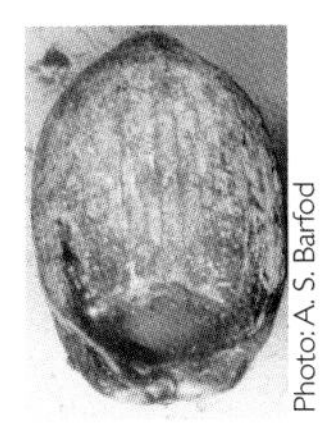

Photo: A. S. Barfod

Borassus heineanus fruit, Sandaun, PNG

Confused with:

- *Corypha*: inflorescences above the leaves, stems dying after flowering, petioles spiny, fruit small.
- *Livistona*: petiole base not cleft, leaf sheaths fibrous, petiole spiny, fruit small.

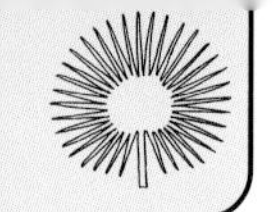

Corypha

Tall – palmate – petiole base cleft – spiny petiole – leaf segments not wedge-shaped

Look for:

- Massive, single-stemmed fan palm with distinctive triangular cleft at base of petiole.
- Petiole with regular, tooth-like spines.
- Inflorescences above the leaves, stem dying after flowering.

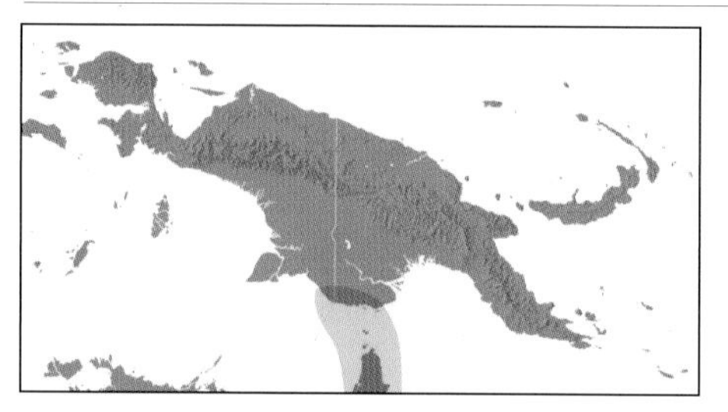

Distribution
India to Philippines, through South-East Asia to Australia, in New Guinea restricted to the central south.

Habitat
Seasonal lowland habitats near to sea level.

Number of species
Approximately 6, of which 1 in New Guinea.

Taxonomic accounts
Checklist in Govaerts & Dransfield 2005.

Uses
Stem for sago.

Habit
Massive, single-stemmed fan palm, height c. 15 m, stem diameter to 90 cm, crownshaft absent, stems dying after flowering, hermaphrodite.

Photo: A. S. Barfod

Corypha utan in flower, Cape York, Australia

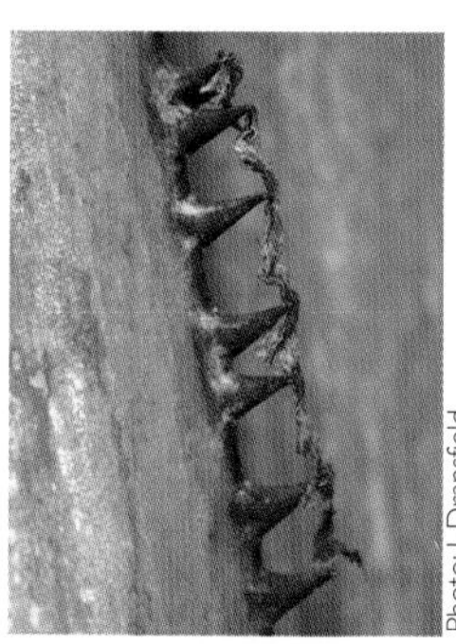

Photo: J. Dransfield

Corypha utan petiole spines, Haad Yai Thailand

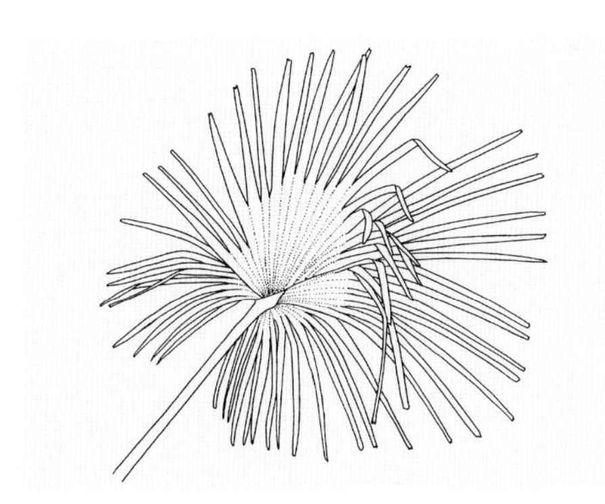

Drawing: P. K. R. Davies

Corypha leaf

Photo: A. S. Barfod

Corypha utan, Cape York, Australia

Leaf palmate, to 500 cm long, to 25 in crown, very large.

Sheath splitting ± to base opposite the petiole, c. 100 cm.

Petiole robust, to 300 cm long, with a conspicuous triangular cleft at the base, the margins with crowded regularly arranged black spines.

Segments at least 80, with pointed tips, the blade split to c. one-third or more of its radius, blade to 300 cm long.

Inflorescence between the leaves, many inflorescences produced simultaneously, individual inflorescences branched to 4 orders, to 300 cm long, branches widely spreading, flowering resulting in the death of the stem.

Primary bracts tubular, tightly sheathing, not dropping off as inflorescence expands.

Peduncle shorter than inflorescence rachis, to 30 cm.

Rachillae very slender, short and straight.

Flowers very small, hermaphrodite, in groups of up to 5.

Fruit green to black, globose, c. 2 cm × c. 2 cm, stigmatic remains basal, flesh rather dry, endocarp thin.

Seed 1, globose, endosperm homogeneous.

Drawing: L.T. Smith

Corypha utan flower

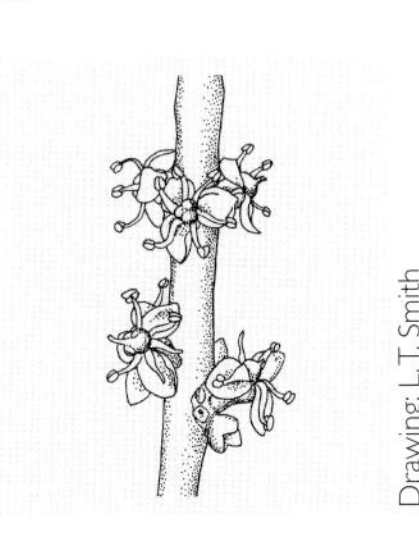

Drawing: L.T. Smith

Corypha utan flowers on rachilla

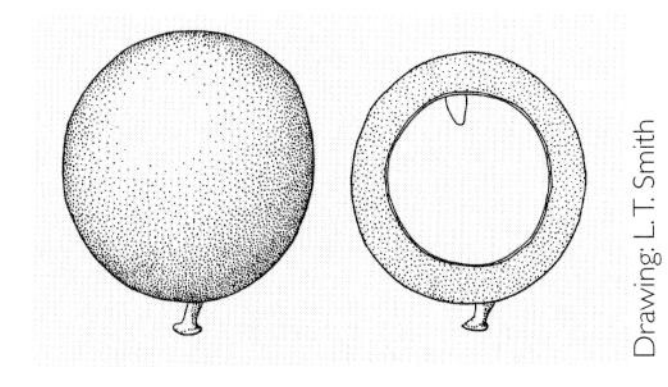

Drawing: L.T. Smith

Corypha utan fruit whole and in section

Confused with:

- *Livistona*: petiole base not cleft, leaf sheaths fibrous, stem not dying after flowering.
- *Borassus*: petiole margins sharp, not spiny, fruit very large (12–15 cm long), stem not dying after flowering.

Licuala

Small to medium – palmate – petiole usually spiny – leaf segments wedge-shaped

Look for:

- Under- to mid-storey, single- or multi-stemmed fan palms.
- Leaf usually split to the base forming wedge-shaped leaflets.
- Inflorescences between the leaves.

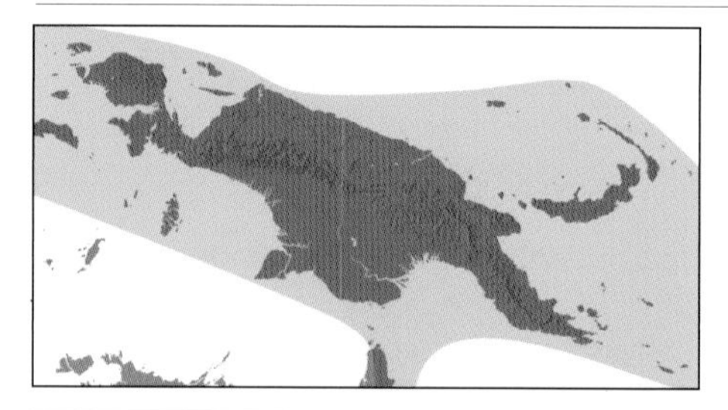

Distribution

South China to India, through South-East Asia to Vanuatu and Australia, widespread in New Guinea.

Habitat

Lowland to montane rainforest, sea level to 1800 m.

Number of species

Approximately 150, of which 27 in New Guinea.

Taxonomic accounts

Barfod 2000, Banka & Barfod 2004, checklist in Govaerts & Dransfield 2005.

Uses

Stem for construction, axe handles, walking sticks, bows and arrows; leaves for thatch.

Habit

Under- to mid-storey, single- or multi-stemmed fan palms, height to 7 m, some species stemless, stem diameter 2.5–11 cm, hermaphrodite.

Photo: W. J. Baker

Licuala lauterbachii, Prafi River, Papua

Photo: A. S. Barfod

Licuala sp. flowers, Alotau, PNG

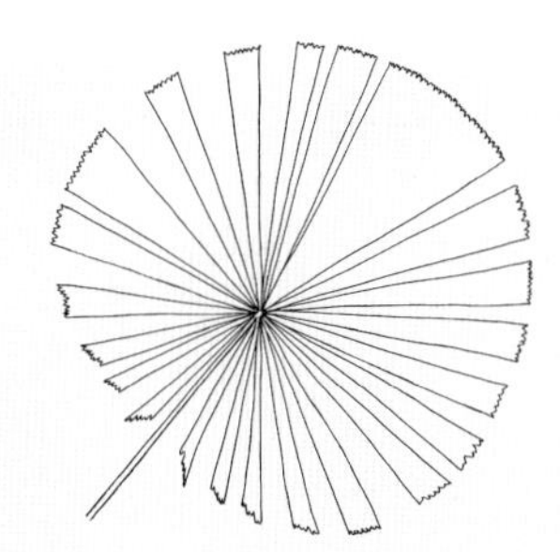

Drawing: P. K. R. Davies

Licuala leaf

Leaf palmate, up to 400 cm long, 10–35 in crown.

Sheath disintegrating into a brown fibrous network, to 35 cm long, not forming crownshaft.

Petiole long, 12–300 cm long, usually with spines along the margins.

Blade divided to base into 2–25 wedge-shaped segments, segments 35–180 cm long, with toothed or lobed tips, sometimes of very varying width, ± horizontal.

Inflorescence between the leaves, unbranched or branched 1–3 orders, 15–450 cm long.

Primary bracts similar, tubular, few to many, not enclosing inflorescence in late bud, not dropping off as inflorescence expands.

Peduncle shorter to longer than inflorescence rachis, 10–80 cm.

Rachillae generally slender, straight or curved.

Flowers generally small, borne singly or in groups of up to 7.

Fruit very variable, red to brown, globose, ovoid or ellipsoid, 0.7–4.5 cm × 0.7–3 cm, stigmatic remains apical, flesh juicy or fibrous, endocarp smooth or heavily ridged and grooved.

Seed 1, endosperm homogeneous, with seed coat intruding into one side of the seed.

Photo: W. J. Baker

Licuala lauterbachii inflorescence, Kikori River, PNG

Photo: W. J. Baker

Licuala lauterbachii leaf, Kikori River, PNG

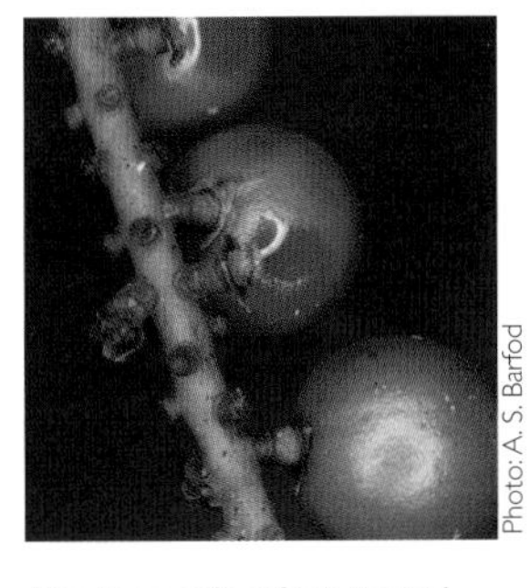

Photo: A. S. Barfod

Licuala parviflora fruit, Bewani, PNG

Confused with:

- Cannot be confused with other fan palms because of the deeply split leaf blade and wedge-shaped segments.

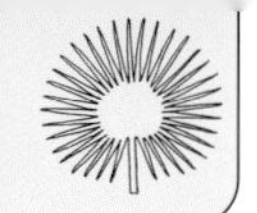

Licuala – some examples

Licuala sp., Tabubil, PNG

Photo: W. J. Baker

Licuala tanycola, Tabubil, PNG

Photo: W. J. Baker

Licuala tanycola inflorescence, Tabubil, PNG

Photo: W. J. Baker

Licuala telifera, Wandammen Peninsula, Papua

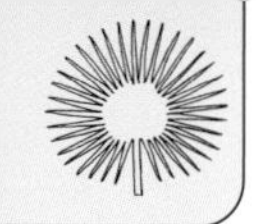

Livistona

Medium to tall – palmate – petiole usually spiny – leaf segments not wedge-shaped

Look for:

- Robust, single-stemmed fan palms with fibrous leaf sheaths.
- Petioles usually spiny.
- Inflorescences between the leaves, highly branched.

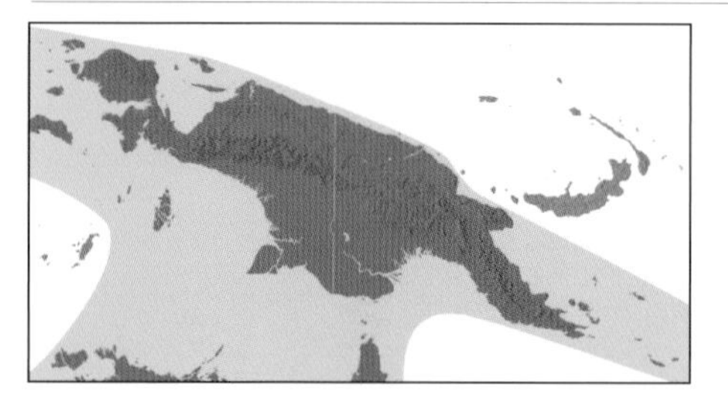

Distribution
Africa and Arabia, South China to India, through South-East Asia to Australia and Solomon Islands, widespread in New Guinea.

Habitat
Lowland to submontane rainforest and seasonal habitats, sea level to 1000 m.

Number of species
Approximately 34, of which 8 in New Guinea.

Taxonomic accounts
Dowe 2001, Dowe & Barfod 2001.

Uses
Stem for bows and floorboards; leaves for thatch and umbrellas; petioles burnt to ash for salt.

Habit
Robust, single-stemmed fan palms, height to 40 m, stem diameter 12–30 cm, hermaphrodite, rarely dioecious.

Photo: W. J. Baker

Livistona surru, Ramu River, PNG

Photo: A. S. Barfod

Livistona surru fruit, Torricelli Mountains, PNG

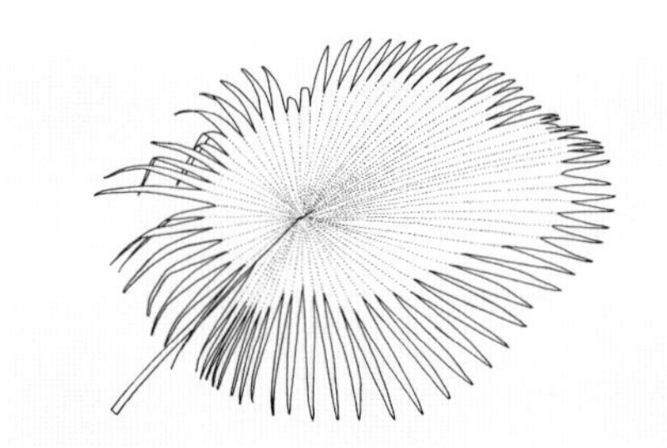

Drawing: P. K. R. Davies

Livistona leaf

Photo: W. J. Baker

Livistona papuana leaf, Timika, Papua

Leaf palmate, to 290 cm long, 16–60 in crown.

Sheath disintegrating into a brown fibrous network, to 40 cm long, not forming crownshaft.

Petiole long, 110–200 cm long, usually with spiny margins.

Segments to 90, with pointed tips, the blade split to c. one-third or more of its radius, blade to 180 cm long.

Inflorescence between the leaves, branched 3–4 orders, 60–225 cm long.

Primary bracts similar, tubular, few to many, not enclosing inflorescence in late bud, not dropping off as inflorescence expands.

Peduncle shorter than inflorescence rachis, 20–40 cm.

Rachillae generally slender, straight or curved.

Flowers generally small, hermaphrodite, borne in groups of up to five, rarely solitary.

Fruit red, brown or black, globose, 0.8–6.5 cm × 0.8–5.5 cm, stigmatic remains apical, flesh juicy or fibrous, endocarp smooth.

Seed 1, endosperm homogeneous, with seed coat intruding into one side of the seed.

Photo: W. J. Baker

Livistona papuana leaf sheaths, Timika, Papua

Photo: A. S. Barfod

Livistona tothur inflorescence, Niau, PNG

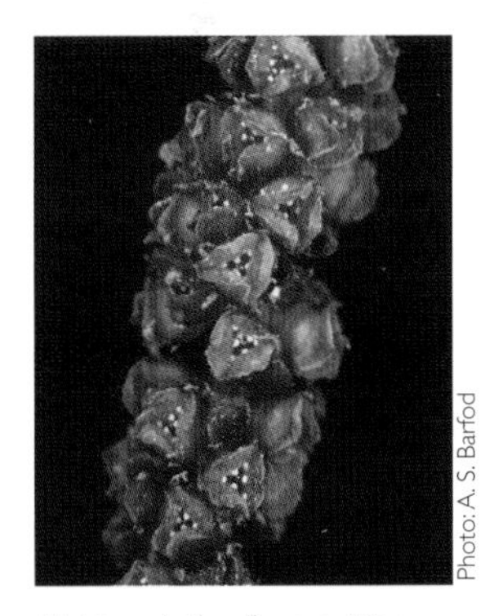

Photo: A. S. Barfod

Livistona tothur flowers, Niau, PNG

Confused with:

- *Borassus*: distinctive triangular cleft at base of petiole, petiole margins sharp, not spiny, fruit very large (12–15 cm long).
- *Corypha*: distinctive triangular cleft at base of petiole, inflorescences above the leaves, stems dying after flowering.

Caryota

Tall – bipinnate – no crownshaft – no spines – leaflets jagged

Look for:

- Robust, single-stemmed tree palm with massive bipinnate leaves and wedge-shaped leaflets.
- Inflorescence below the leaves, oldest inflorescences towards top of stem.
- Stems dying after flowering.

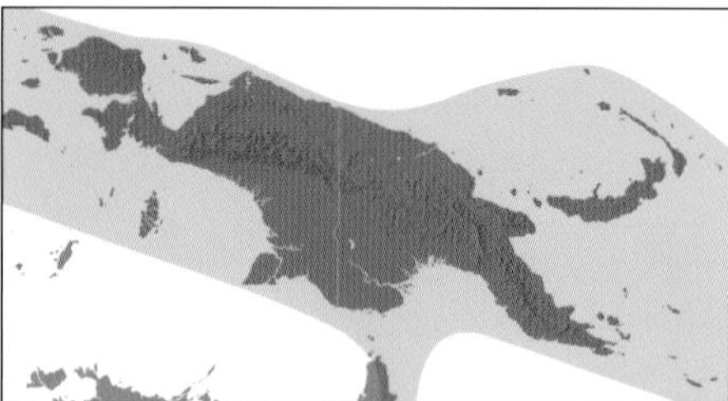

Distribution
South China to India, through South-East Asia to Australia, widespread in New Guinea.

Habitat
Lowland to montane rainforest, sea level to 1500 m.

Number of species
13, of which 2 in New Guinea.

Taxonomic accounts
Dransfield et al. 2000.

Uses
Stem for axe handles, bows and arrows, construction and sago; leaf rachis for fishing rods; fruits eaten after prolonged boiling; seed sometimes used as a betel substitute.

Habit
Robust, single-stemmed tree palm, height to 20 m, stem diameter 19–30 cm, no crownshaft, stems dying after flowering, monoecious.

Photo: W.J. Baker

Caryota rumphiana, Madang, PNG

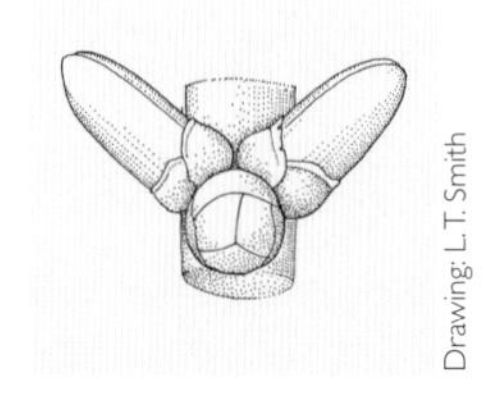
Drawing: L. T. Smith

Caryota rumphiana flower buds

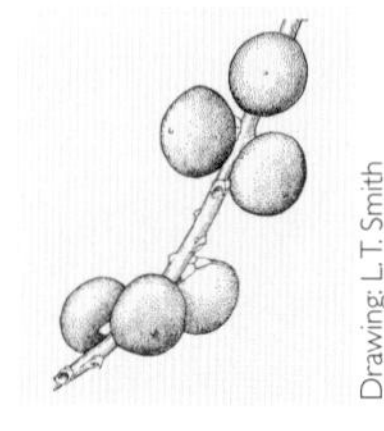
Drawing: L. T. Smith

Caryota rumphiana fruit

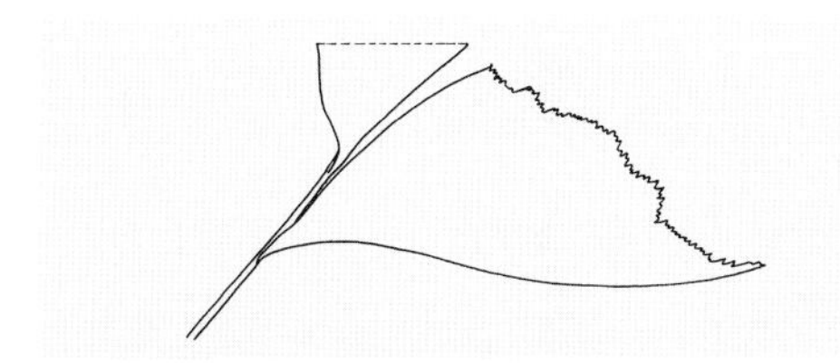
Drawing: P. K. R. Davies

Caryota leaflet

Photo: W. J. Baker

Caryota zebrina leaf sheaths and petioles, private collection, Bogor, Indonesia

Photo: W. J. Baker

Caryota rumphiana inflorescence, Mubi River, PNG

Leaf bipinnate, to 500 cm long, 7–8 in crown, ± horizontal.

Sheath to 150 cm long, disintegrating into a mass of dark fibres, crownshaft absent.

Petiole short to long, 4–150 cm long (longer in juveniles).

Leaflets c. 25 each side of secondary leaf rachis, wedge-shaped, 8–12 cm long, with jagged tips, arranged regularly, horizontal or on edge, V-shaped in section at the base (induplicate).

Inflorescence between the leaves, maturing from top of stem downwards, thus the oldest inflorescence at the stem tip, branched 1–3 orders, to 200 cm long, branches curved, ± pendulous.

Prophyll small, inconspicuous, peduncular bracts 5–7, soon splitting, densely covered in indumentum, not dropping off as inflorescence expands.

Peduncle shorter than or about the same length as rachis, 55–75 cm.

Rachillae straight, pendulous.

Flowers

Flowers in triads throughout the length of the rachilla, not developing in pits, male flowers ± bullet-shaped.

Fruit

Fruit red or black, 1.5–2 cm × 1.5–2 cm, stigmatic remains apical, flesh juicy, filled with irritant needle crystals.

Seeds 1 or 2, ± globose or hemispherical, endosperm homogeneous or ruminate.

Confused with:

- *Arenga*: leaves pinnate (not bipinnate), leaflets linear.

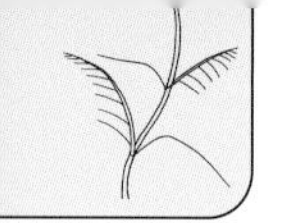

Calamus

Climber – pinnate – spiny – leaflets not diamond-shaped – leaflets pointed

Look for:

- Climbing rattan palms, single- or multi-stemmed, with spiny leaves and leaf sheaths.
- Climbing whips arising from leaf tips or leaf sheaths.
- Primary bracts of inflorescence tubular (at least at the base).

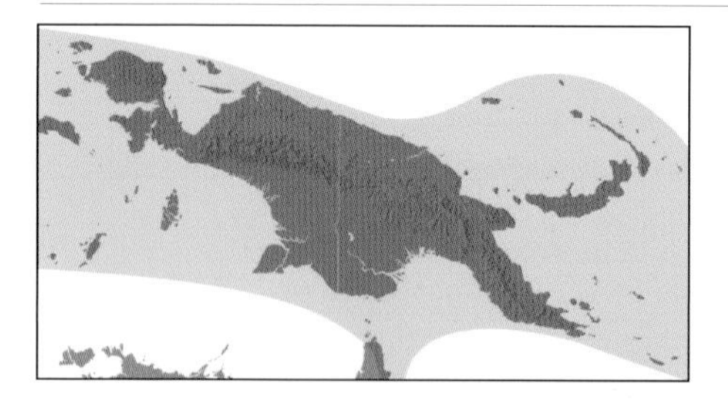

Calamus aruensis, Kikori, PNG

Distribution
Africa, South China to India, through South-East Asia to Fiji and Australia, widespread in New Guinea.

Habitat
Lowland to montane rainforest, sea level to 2800 m.

Number of species
Approximately 370, of which around 60 in New Guinea.

Taxonomic accounts
Baker 2002b, Baker & Dransfield 2002a, 2002b, Baker et al. 2003, Dransfield & Baker 2003, checklist in Govaerts & Dransfield 2005.

Uses
Cane (stem) widely used, e.g. binding in construction, bridge cables, furniture, weaving baskets, armbands and other handicrafts, waist bands, fire-making straps, bow strings; leaves for wrapping food; cirrus for catching fish and flying foxes; fruit edible; sap said to have medicinal properties.

Habit
Slender to robust, single- or multi-stemmed climbing palms, height to 50 m, stem diameter (including leaf sheaths) 0.4–4.5 cm, dioecious.

Calamus sp. male flowers, Kuala Kencana, Papua

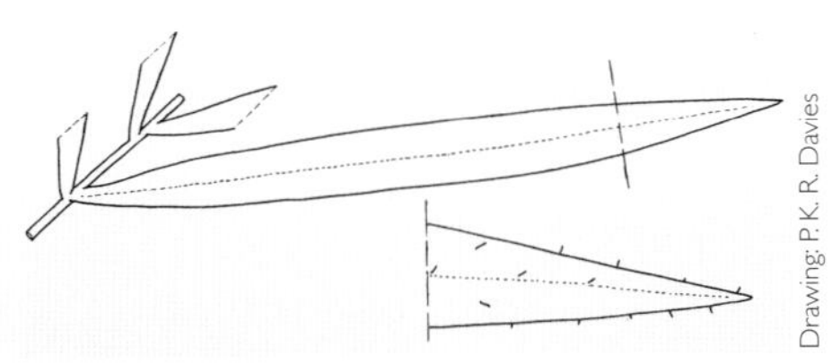
Drawing: P. K. R. Davies

Calamus leaflet

Photo: W.J. Baker

Calamus aruensis leaf sheath with inflorescence, Ramu River, PNG

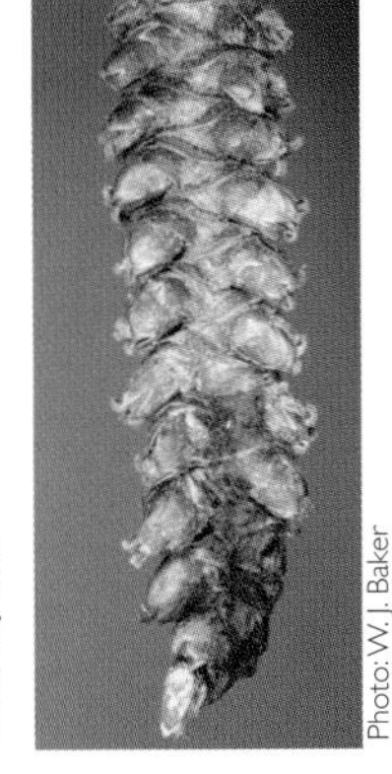
Photo: W.J. Baker

Calamus eximius female flowers, near Tabubil, PNG

Photo: W.J. Baker

Calamus aruensis female inflorescence, Timika, Papua

Photo: W.J. Baker

Calamus longipinna fruit, Ramu River, PNG

Leaf pinnate, 12–500 cm long (including cirrus if present), armed with spines and bristles, ± straight to arching.

Sheath tubular, usually very spiny.

Petiole 0–40 cm long.

Leaflets 3–90 each side of leaf rachis, 12–56 cm long, with pointed tips, arranged regularly or irregularly, horizontal to pendulous.

Spiny climbing whips arising from leaf tips (cirrus) or leaf sheaths (flagellum).

Inflorescence between the leaves, branched 1–3 orders, 10–600 cm long, erect, pendulous or whip-like.

Prophyll and peduncular bracts similar, not enclosing inflorescence in late bud, not dropping off as inflorescence expands, remaining tubular, at least at the base, often spiny.

Peduncle usually shorter than inflorescence rachis, 1.5–380 cm.

Rachillae slender, straight or curved.

Flowers

Flowers in pairs (female inflorescence) or solitary (male inflorescence) throughout the length of the rachilla, not developing in pits.

Fruit

Fruit various colours, globose to ellipsoid, 1–3 cm × 0.8–1.5 cm, with vertical rows of scales, stigmatic remains apical, flesh thin.

Seed 1, enclosed in a fleshy coat (sarcotesta), various shapes, endosperm homogeneous or ruminate.

Confused with:

- *Daemonorops*: primary bracts splitting to base and falling.
- *Korthalsia*: diamond-shaped leaflets, stems dying after flowering.
- *Metroxylon* (as stemless juvenile): leaf rachis spines not grapnel like.
- *Pigafetta* (as stemless juvenile): leaf rachis spines not grapnel like, leaf sheaths chalky white.

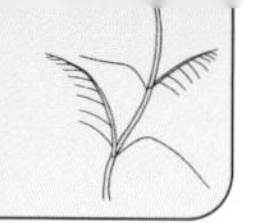

Calamus – some examples

Calamus humboldtianus, near Finschhafen, PNG

Calamus humboldtianus inflorescence in fruit, Timika, Papua

Calamus longipinna, near Lae, PNG

Calamus longipinna leaf sheath, near Lae, PNG

Photo: W. J. Baker

Calamus warburgii male inflorescence, near Madang, PNG

Photo: W. J. Baker

Calamus warburgii, Kikori, PNG

Photo: W. J. Baker

Calamus fertilis, Mubi River, PNG

Photo: W. J. Baker

Calamus fertilis, Kikori, PNG

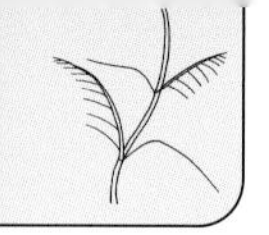

Daemonorops

Climber – pinnate – spiny – leaflets not diamond-shaped – leaflets pointed

Look for:

- Climbing, multi-stemmed rattan palm with spiny leaves and leaf sheaths.
- Climbing whips arising from leaf tips.
- Primary bracts very spiny, splitting to base and usually dropping off.

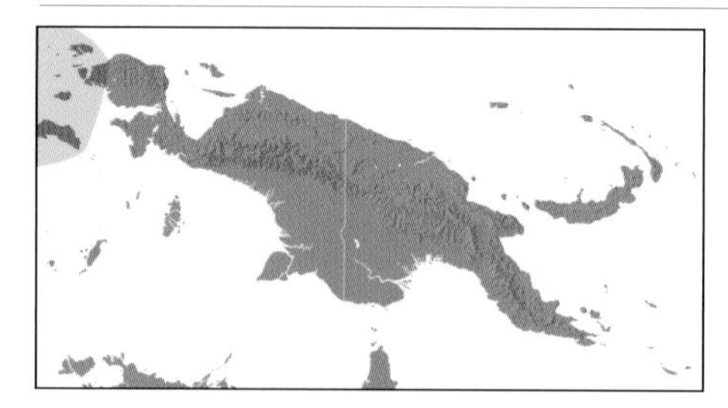

Distribution
South China to India, through South-East Asia to New Guinea, in New Guinea restricted to the far west.

Habitat
Lowland rainforest, sea level to 120 m.

Number of species
Approximately 100, of which 1 in New Guinea.

Taxonomic accounts
Maturbongs 2003.

Uses
Leaves for thatch.

Habit
Robust, multi-stemmed climbing palm, height 25 m, stem diameter (including leaf sheaths) 4.5–7 cm, dioecious.

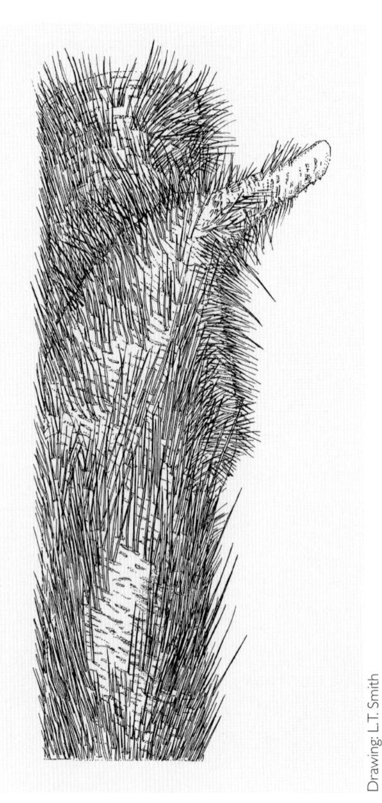

Drawing: L.T. Smith

Daemonorops sp. leaf sheath

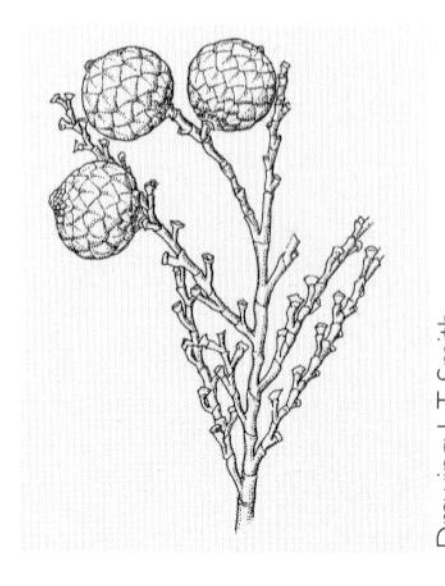

Drawing: L.T. Smith

Daemonorops sp. part of inflorescence with fruit

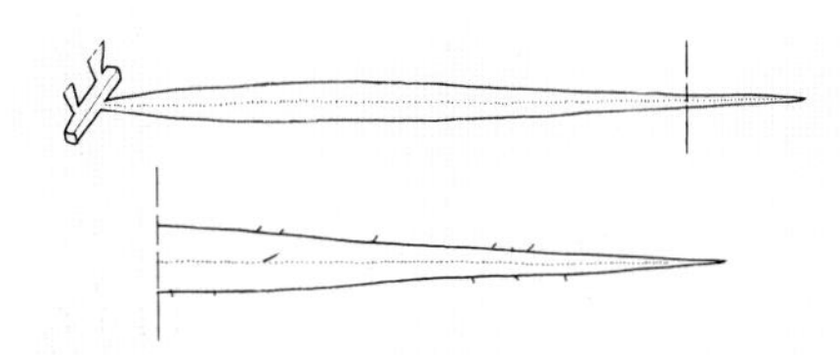

Daemonorops leaflet

Daemonorops sp. inflorescence

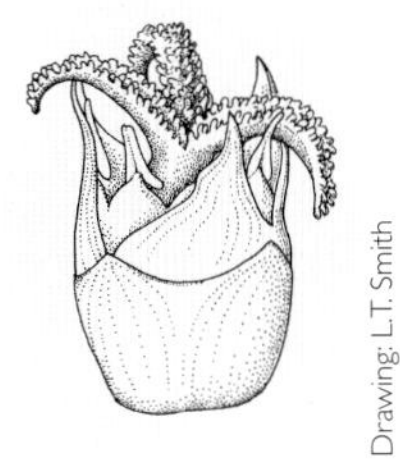

Daemonorops sp.
female flower

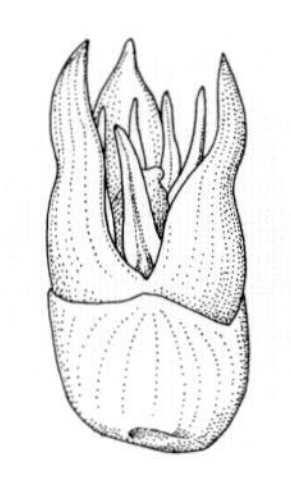

Daemonorops sp.
male flower

Leaf pinnate, 400–600 cm long (including cirrus), armed with spines and bristles, arching.

Sheath tubular, usually very spiny.

Petiole 30–55 cm long.

Leaflets 68–91 each side of leaf rachis, to 65 cm long, with pointed tips, arranged regularly, ± horizontal.

Spiny climbing whips arising from leaf tips (cirrus).

Inflorescence between the leaves, branched 2–3 orders, 60–120 cm long, pendulous.

Primary bracts similar, not enclosing inflorescence in late bud, splitting and usually dropping off as inflorescence expands, very spiny.

Peduncle shorter than inflorescence rachis, c. 20 cm.

Rachillae slender, straight.

Flowers in pairs (female inflorescence) or solitary (male inflorescence) throughout the length of the rachilla, not developing in pits.

Fruit yellow-brown, globose, 2–2.5 cm × 1.5–2 cm, with vertical rows of scales, stigmatic remains apical, flesh thin.

Seed 1, enclosed in a fleshy coat (sarcotesta), globose, endosperm ruminate.

Confused with:

- *Calamus*: climbing whips arising from leaf sheaths (flagellum) in many species (but also from leaf tips in other species), primary bracts of inflorescence tubular (at least at the base).
- *Korthalsia*: diamond-shaped leaflets, stems dying after flowering.
- *Metroxylon* (as stemless juvenile): leaf rachis spines not grapnel like.
- *Pigafetta* (as stemless juvenile): leaf rachis spines not grapnel like, leaf sheaths chalky white.

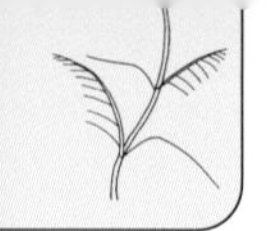

Korthalsia

Climber – pinnate – spiny – leaflets diamond-shaped – leaflets jagged

Look for:

- Climbing, multi-stemmed rattan palm with spiny leaves and leaf sheaths, and diamond-shaped leaflets.
- Climbing whips arising from leaf tips.
- Stem dying after flowering.

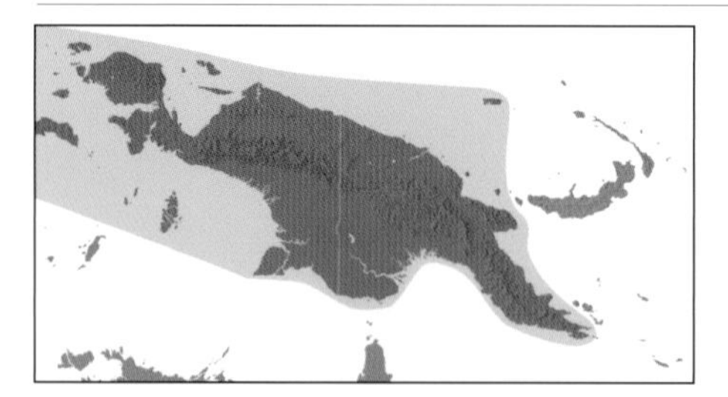

Distribution
India to Vietnam, through South-East Asia to New Guinea, widespread in New Guinea.

Habitat
Lowland rainforest, sea level to 1100 m.

Number of species
Approximately 27, of which 1 in New Guinea.

Taxonomic accounts
Dransfield 1981.

Uses
Cane (stem) for matting, binding houses; cirrus in fish traps.

Habit
Moderately robust, multi-stemmed climbing palm, height to 25 m, stem diameter (including leaf sheaths) 2–4 cm, stems dying after flowering, hermaphrodite.

Photo: W. J. Baker

Korthalsia zippelii, Timika, Papua

Photo: W. J. Baker

Korthalsia zippelii rachillae, Madang, PNG

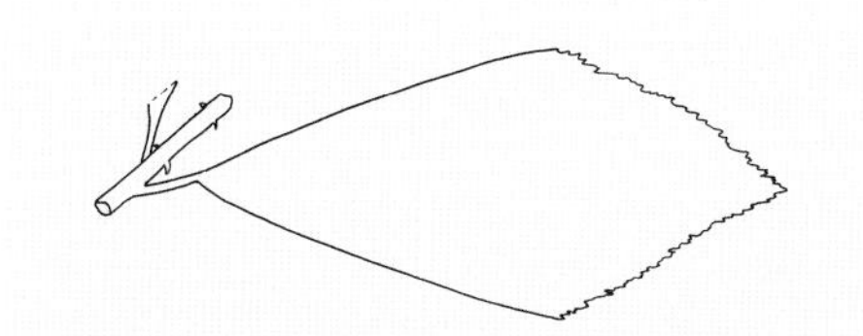

Drawing: P. K. R. Davies

Korthalsia leaflet

Photo: J. Dransfield

Korthalsia zippelii fruit, Manokwari, Papua

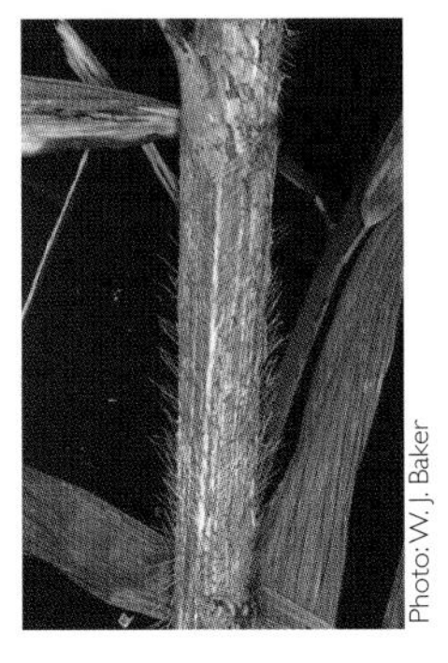

Photo: W. J. Baker

Korthalsia zippelii leaf sheath, near Lae, PNG

Photo: W. J. Baker

Korthalsia zippelii stem apex with inflorescences, near Lae, PNG

Leaf pinnate, 2–3 m long (including cirrus), armed with spines and bristles, arching.

Sheath tubular, usually very spiny.

Petiole 8–60 cm long.

Leaflets 9–12 each side of leaf rachis, 20–42 cm long, diamond-shaped, with jagged tips, arranged regularly, pendulous.

Spiny climbing whips arising from leaf tips (cirrus).

Inflorescence between the leaves, branched to 2 orders, 30–100 cm long, erect to arching.

Primary bracts similar, not enclosing inflorescence in late bud, not dropping off as inflorescence expands, remaining tubular, usually not spiny.

Peduncle shorter than inflorescence rachis, 2–10 cm.

Rachillae robust, sausage-like, straight or curved, with conspicuous bracts.

Flowers solitary throughout the length of the rachilla, developing in pits formed by rachilla bracts.

Fruit orange-brown, ovoid, 0.8–1.2 cm × 0.8–1 cm, with vertical rows of scales, stigmatic remains apical, flesh thick and pulpy.

Seed 1, not enclosed in a fleshy coat (sarcotesta), ellipsoid, endosperm homogeneous.

Confused with:

- *Calamus*: climbing whips arising from leaf sheaths (flagellum) in many species (but also from leaf tips (cirrus) in other species), leaflets not diamond-shaped, stems not dying after flowering.
- *Daemonorops*: leaflets not diamond-shaped, stems not dying after flowering.
- *Sommieria*: not climbing, leaflets not diamond-shaped, not spiny, can only be confused with juvenile *Korthalsia*.

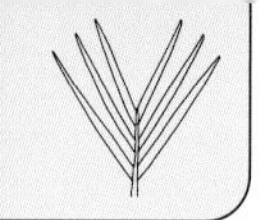

Actinorhytis

Tall – pinnate – crownshaft – no spines – leaflets pointed

Look for:

- Robust, single-stemmed tree palm with strongly arching leaves and slender crownshaft.
- Inflorescence below the leaves, widely spreading and branched to 3 orders.
- Fruit 6 cm or more long, seed with endosperm strongly ruminate.

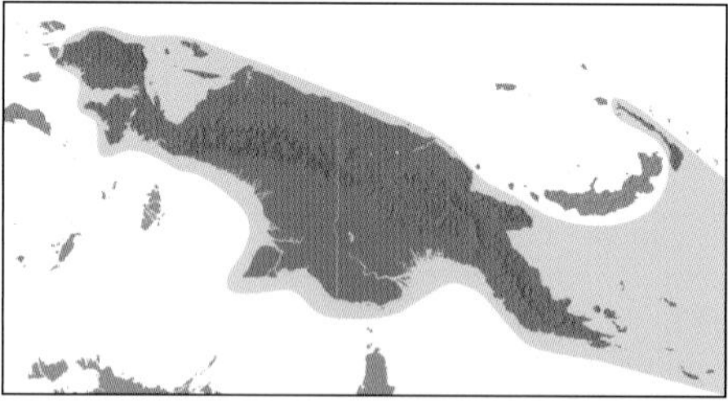

Distribution
New Guinea to Solomon Islands, frequently cultivated in South-East Asia.

Habitat
Lowland to submontane rainforest, sea level to 1000 m.

Number of species
1 species.

Taxonomic accounts
Checklist in Govaerts & Dransfield 2005.

Uses
Stem for bows; seed as a substitute for betel.

Habit
Robust, single-stemmed tree palm, height to 30 m, stem diameter 13–30 cm, slender crownshaft ± the same diameter as stem, monoecious.

Actinorhytis calapparia, Lae Botanic Garden, PNG

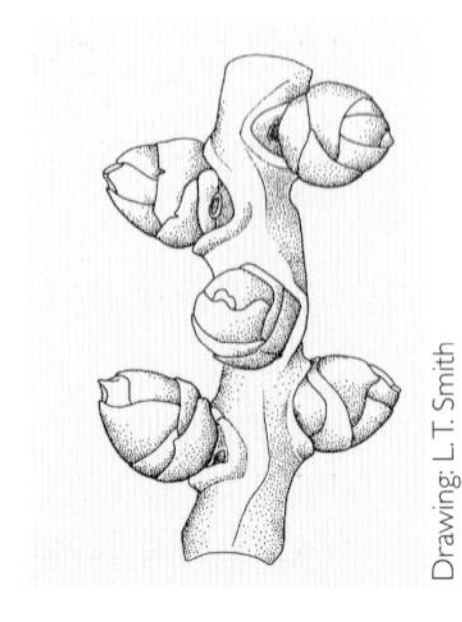

Actinorhytis calapparia rachilla with female flowers

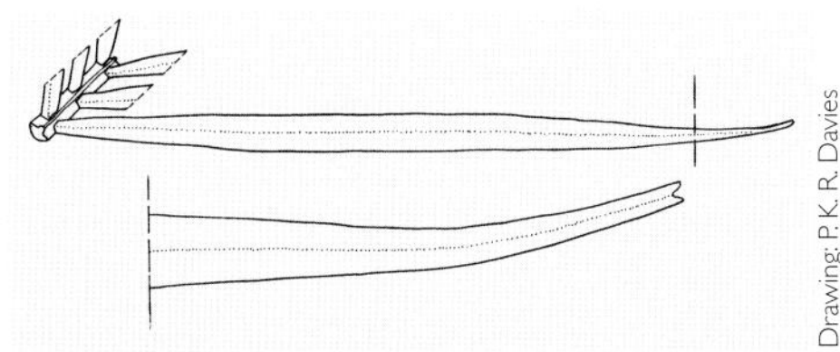

Drawing: P. K. R. Davies

Actinorhytis leaflet

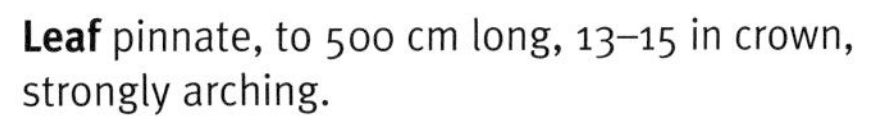

Leaf pinnate, to 500 cm long, 13–15 in crown, strongly arching.

Sheath tubular, pale green, to 200 cm long, forming crownshaft.

Petiole short, 20–50 cm long.

Leaflets 55–110 each side of leaf rachis, to 100 cm long, with pointed tips, arranged regularly, ascending.

Inflorescence below the leaves, branched to 3 orders, to 75 cm long, branches widely spreading.

Prophyll and peduncular bract similar, enclosing inflorescence in bud, dropping off as inflorescence expands.

Peduncle shorter than inflorescence rachis, 15–25 cm, grossly swollen at base.

Rachillae slender and straight.

Flowers in triads at the base of the branches, pairs of male flowers towards tip, not developing in pits, female flowers much larger than the male flowers.

Fruit red, ovoid, 6–8 cm × 4–5 cm, stigmatic remains apical, flesh fibrous, endocarp thin, closely adhering to seed.

Seed 1, globose, endosperm strongly ruminate.

Photo: W. J. Baker

Actinorhytis calapparia, Lae Botanic Garden, PNG

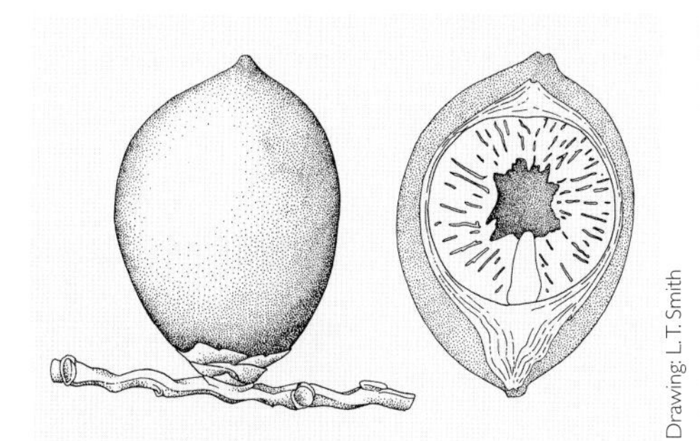

Drawing: L. T. Smith

Actinorhytis calapparia fruit whole and in section

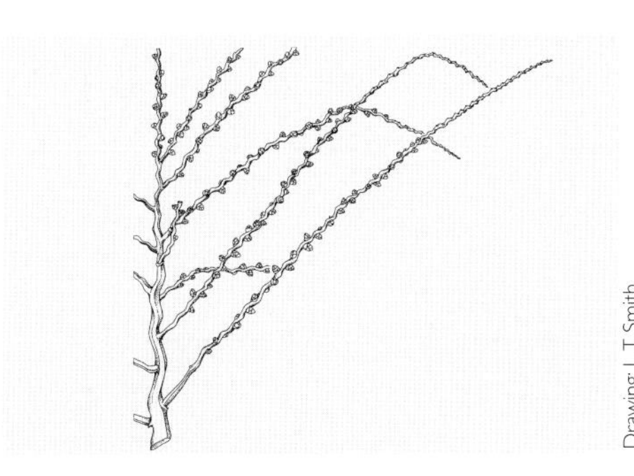

Drawing: L. T. Smith

Actinorhytis calapparia inflorescence portion

Confused with:

- *Hydriastele*: jagged leaflet tip, horse's tail-like inflorescence, small fruit.

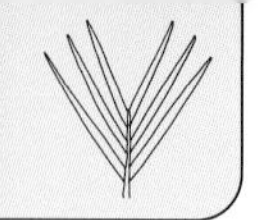

Areca

Small to medium – pinnate – crownshaft – no spines – leaflets lobed or pointed

Look for:

- Understorey to mid-storey, single-stemmed tree palms.
- Inflorescence below the leaves, female flowers occurring only at the base of the branches.
- Fruiting inflorescence often club-like.

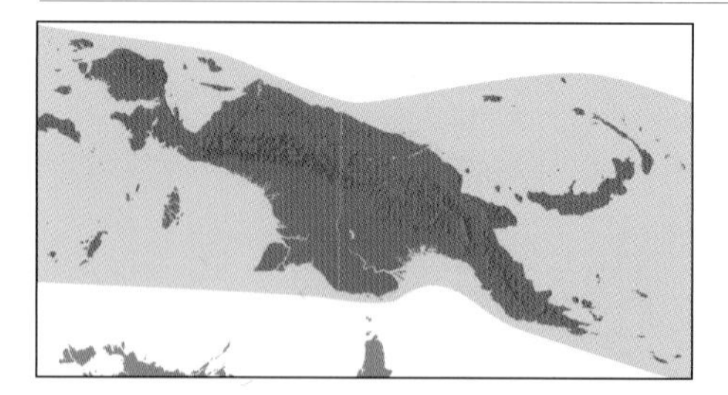

Distribution
South China to India, through South-East Asia to Solomon Islands, widespread in New Guinea.

Habitat
Lowland to montane rainforest, sea level to 1800 m.

Number of species
Approximately 48, of which 3 or more in New Guinea.

Taxonomic accounts
Flynn 2004.

Uses
Leaves for thatch; seed for betel.

Habit
Single-stemmed tree palms, height to 15 m, stem diameter 2–12 cm, crownshaft present, variously coloured, monoecious.

Photo: W. J. Baker

Areca macrocalyx, Kikori, PNG

Photo: W. J. Baker

Areca macrocalyx fruit, Kikori, PNG

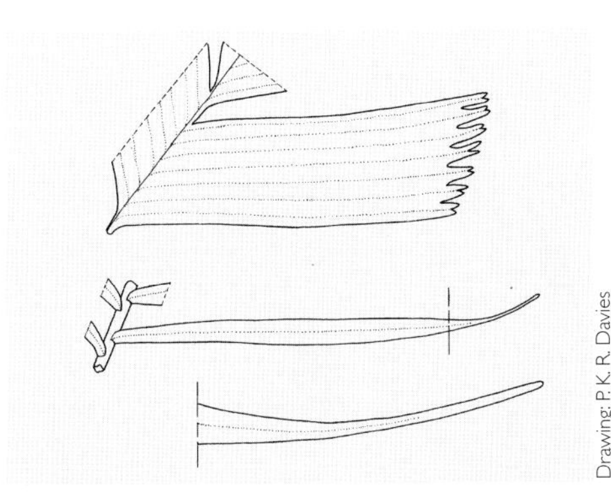
Drawing: P. K. R. Davies

Areca leaflet variation

Photo: W. J. Baker

Areca macrocalyx (red form) flowers, near Finschhafen, PNG

Photo: W. J. Baker

Areca macrocalyx inflorescences in flower and fruit, Kikori, PNG

Photo: W. J. Baker

Areca macrocalyx inflorescence, near Tabubil, PNG

Leaf pinnate, 300 cm long, 5–8 in crown, straight.

Sheath tubular, variously coloured, 25–150 cm long, forming crownshaft.

Petiole 2–40 cm long.

Leaflets 5–40 each side of leaf rachis, rarely single-fold, usually composed of many folds, to 75 cm long, with pointed or lobed tips, arranged regularly or irregularly, ± horizontal.

Inflorescence below the leaves, branched 2–3 orders, to 60 cm long, branches widely spreading.

Prophyll thin, enclosing inflorescence in bud, dropping off as inflorescence expands, peduncular bract absent.

Peduncle shorter than inflorescence rachis, 3–12 cm.

Rachillae slender and usually straight.

Flowers in triads at the base of the branches, pairs of male flowers towards tip, not developing in pits, female flowers much larger than the male flowers.

Fruit orange to red, 1.5–4.5 cm × 1–3.5 cm, stigmatic remains apical, flesh fibrous, endocarp thin, closely adhering to seed. Seed 1, globose or ellipsoid, endosperm strongly ruminate.

Confused with:

- *Pinanga*: female flowers and fruit throughout entire length of inflorescence branches.
- *Physokentia*: inflorescence with prophyll and peduncular bract, prophyll not encircling peduncle completely, endocarp ridged.

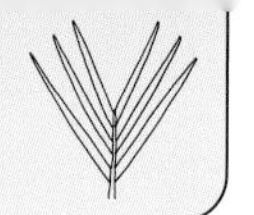

Areca – some examples

Areca macrocalyx (red form), near Finschhafen, PNG

Areca macrocalyx (red form) inflorescence, near Finschhafen, PNG

Areca sp. inflorescence in fruit, Manokwari, Papua

Areca sp., Wandammen Peninsula, Papua

Areca catechu, Malaysia

Areca catechu being sold at market in Mumeng, PNG

Photo: W. J. Baker

Areca novohibernica stilt roots, Lae Botanic Garden, PNG

Areca novohibernica inflorescence, Lae Botanic Garden, PNG

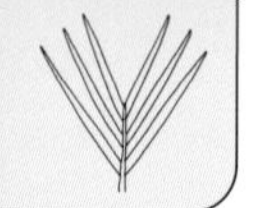

Arenga

Medium – pinnate – no crownshaft – no spines – leaflets jagged

Look for:

- Untidy, multi-stemmed, mid-storey tree palm with conspicuous leaf sheath fibres.
- Leaflet bases asymmetrical and V-shaped in section.
- Stem dying after flowering.

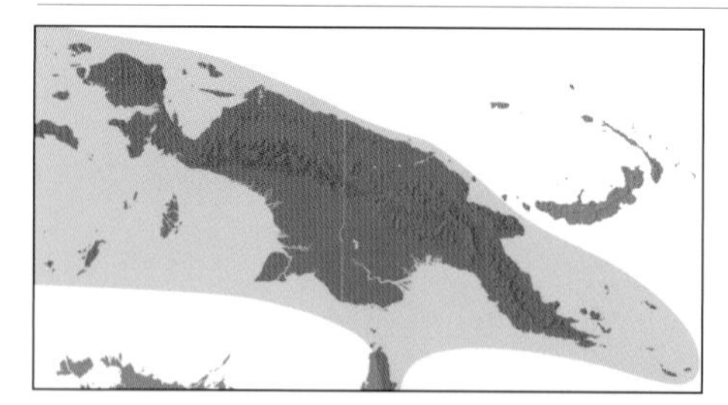

Distribution
South China to India, through South-East Asia to Australia, widespread in New Guinea, not recorded from Bismarck Archipelago.

Habitat
Lowland rainforest, sea level to 700 m.

Number of species
Approximately 24, of which 1 in New Guinea, a second species (*A. pinnata*) cultivated occasionally.

Taxonomic accounts
Mogea 1991.

Uses
Stem for construction and sago; leaf indumentum for tinder; leaves for making skirts; shoot apex edible; leaves fibre for grass skirts.

Habit
Untidy, multi-stemmed, mid-storey tree palm, height to 14 m, stem diameter 10–15 cm, crownshaft lacking, stems dying after flowering, monoecious.

Photo: W.J. Baker

Arenga microcarpa, Kikori, PNG

Photo: W.J. Baker

Arenga microcarpa young fruit, Madang, PNG

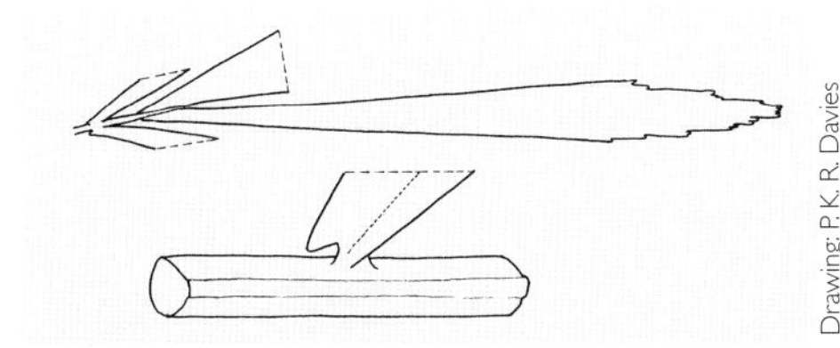

Drawing: P. K. R. Davies

Arenga leaflet

Photo: W. J. Baker

Arenga microcarpa young inflorescence, Madang, PNG

Photo: W. J. Baker

Arenga microcarpa male flowers, Wandammen Peninsula, Papua

Photo: W. J. Baker

Arenga microcarpa inflorescence, Wandammen Peninsula, Papua

Leaf pinnate, to 400 cm long, 5–10 in crown, ± straight.

Sheath to 130 cm long, disintegrating into a mass of fibres.

Petiole long, to 100 cm long.

Leaflets 40–70 each side of rachis, to 110 cm long, with jagged tips, arranged ± regularly, horizontal, undersurface covered with grey indumentum, leaflet bases asymmetrical and V-shaped in section (induplicate).

Inflorescence between the leaves, maturing from top of stem downwards, thus the oldest inflorescence at the stem tip, branched to 1 order, to 110 cm long, branches curved, ± pendulous.

Prophyll very small, inconspicuous, peduncular bracts 5–7, splitting, densely covered in indumentum, not dropping off as inflorescence expands.

Peduncle about the same length as inflorescence rachis, to 30 cm long.

Rachillae quite robust and straight.

Flowers male and female flowers borne in separate inflorescences on the same stem, the female inflorescence usually borne at the stem tip, flowers not developing in pits, male flowers bullet-shaped.

Fruit red, globose, c. 1.5 cm × c. 1.5 cm, stigmatic remains apical, flesh juicy, filled with irritant needle crystals.

Seed 1–3, ± globose or angled, endosperm homogeneous.

Confused with:

- *Caryota*: bipinnate leaves, leaflets wedge-shaped.
- *Orania*: leaflets reduplicate (Λ-shaped), stems not dying after flowering, fruit quite large.

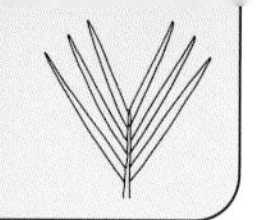

Brassiophoenix

Medium – pinnate – crownshaft – no spines – leaflets jagged

Look for:

- Slender, single-stemmed, mid-storey tree palm with broad wedge-shaped leaflets.
- Inflorescence below the leaves.
- Orange, medium-sized fruit (c. 3.5 cm long) with pale, hard coat (endocarp) enclosing seed with conspicuous grooves and ridges.

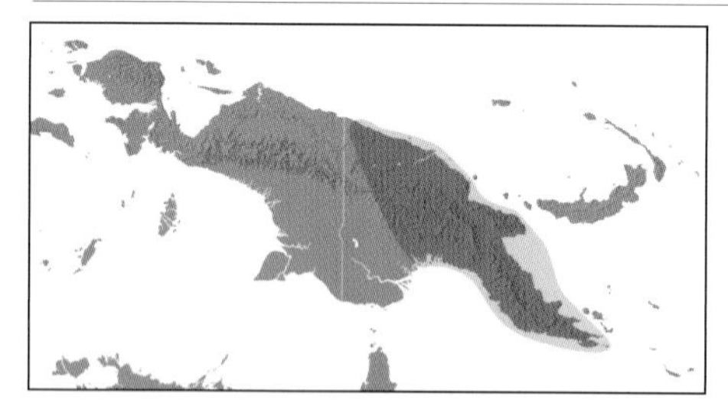

Distribution
Endemic to PNG, not recorded from Bismarck Archipelago.

Habitat
Lowland rainforest, sea level to 800 m.

Number of species
2 species.

Taxonomic accounts
Zona & Essig 1999.

Uses
Seed as a betel substitute.

Habit
Slender, single-stemmed, tree palm, height to 10 m, stem diameter 5–7 cm, crownshaft present, monoecious.

Photo: A. Kjaer

Brassiophoenix schumannii, East Sepik, PNG

Photo: A. S. Barfod

Brassiophoenix schumannii male flowers, Finschhafen, PNG

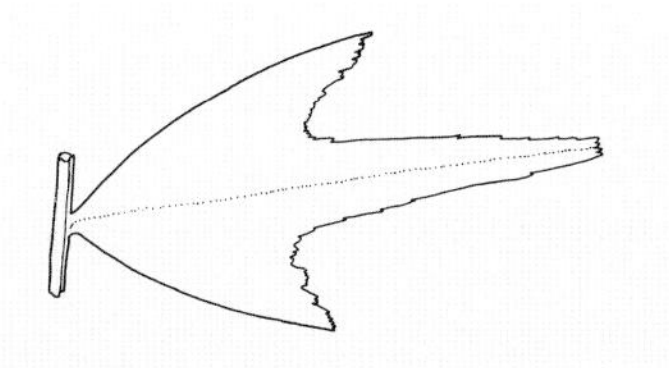

Drawing: P. K. R. Davies

Brassiophoenix leaflet

Photo: W. J. Baker

Brassiophoenix drymophloeoides leaf, Kikori, PNG

Leaf pinnate, 140–230 cm long, 5–12 in crown, straight.

Sheath tubular, 30–60 cm long, forming crownshaft.

Petiole 4–22 cm long.

Leaflets 7–14 each side of leaf rachis, 33–84 cm long, distinctively wedge-shaped leaflets with a broad, jagged apical margin formed into 2 or 3 prongs, arranged regularly, horizontal.

Inflorescence below the leaves, branched 2–3 orders, 25–95 cm long, branches spreading.

Peduncular bract longer than and projecting from prophyll, enclosing inflorescence in bud, prophyll and peduncular bract dropping off as inflorescence expands.

Peduncle shorter or longer than rachis, 11–14 cm.

Rachillae slender and ± curved.

Flowers in triads throughout the length of the rachillae, not developing in pits, male flowers larger than female flowers, bullet-shaped.

Fruit orange, globose, c. 3.5 cm × 3–2.5 cm, stigmatic remains apical, flesh thick and juicy, endocarp pale, thick with conspicuous grooves and ridges, closely adhering to seed.

Seed 1, conforming to endocarp shape, endosperm homogeneous.

Photo: W. J. Baker

Brassiophoenix schumannii inflorescence in fruit, Madang, PNG

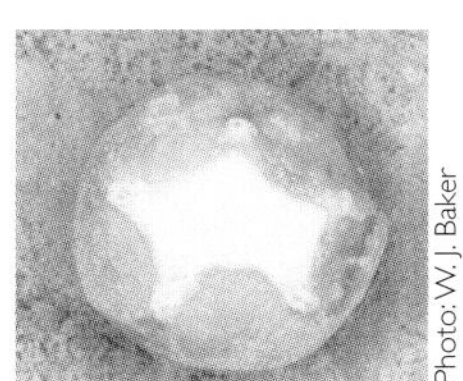

Photo: W. J. Baker

Brassiophoenix drymophloeoides fruit in cross section, Kikori, PNG

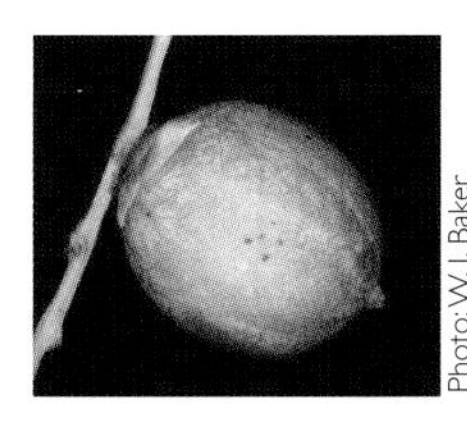

Photo: W. J. Baker

Brassiophoenix drymophloeoides fruit, Kikori, PNG

Confused with:

- *Drymophloeus*: stilt roots, endocarps lacking conspicuous ridges.
- *Ptychococcus*: moderate to robust palm, leaflets linear, fruit red, endocarp black or dark brown.
- *Ptychosperma*: fruit small, endocarps with shallow ridges and grooves.

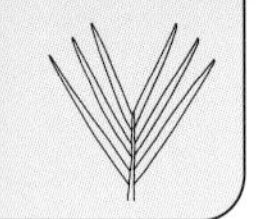

Calyptrocalyx

Very small to medium – pinnate – no crownshaft – no spines – leaflets lobed, toothed or pointed

Look for:

- Understorey (rarely mid-storey), single- or multi-stemmed palms with fibrous leaf sheaths.
- Inflorescence between or below the leaves, unbranched (spicate), peduncular bract arising near base of peduncle, not dropping off.
- Flowers developing within pits in the inflorescence spike.

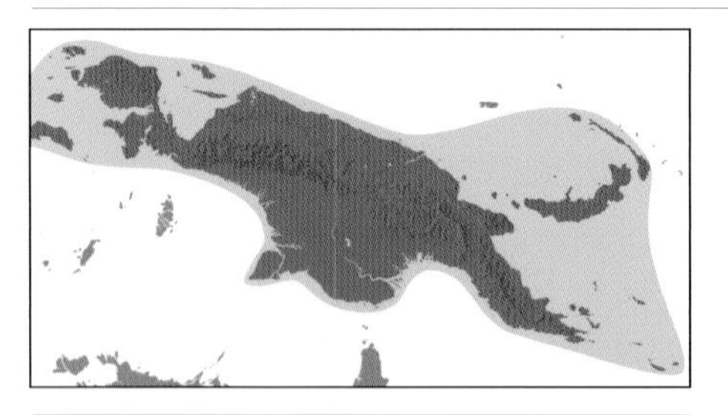

Distribution
Moluccas to New Guinea, widespread in New Guinea.

Habitat
Lowland to submontane rainforest, sea level to 2000 m.

Number of species
Approximately 26, of which 25 in New Guinea.

Taxonomic accounts
Dowe & Ferrero 2001.

Uses
Stems for spears, practice/children's bows, construction and breeding edible larvae; leaves for wrapping food; shoot apex edible; seeds edible, also as betel substitute.

Habit
Slender to moderate, single- or multi-stemmed tree palms, height to 15 m, though usually much smaller, stem diameter 0.4–10 cm, no crownshaft, monoecious.

Photo: W. J. Baker

Calyptrocalyx hollrungii, near Finschhafen, PNG

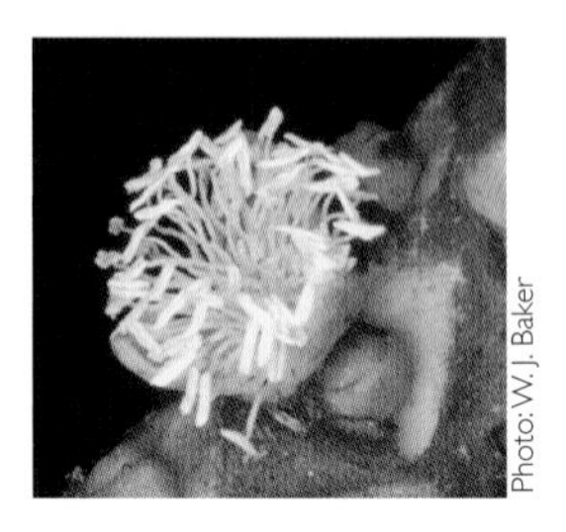

Photo: W. J. Baker

Calyptrocalyx albertisianus male flowers, Kikori, PNG

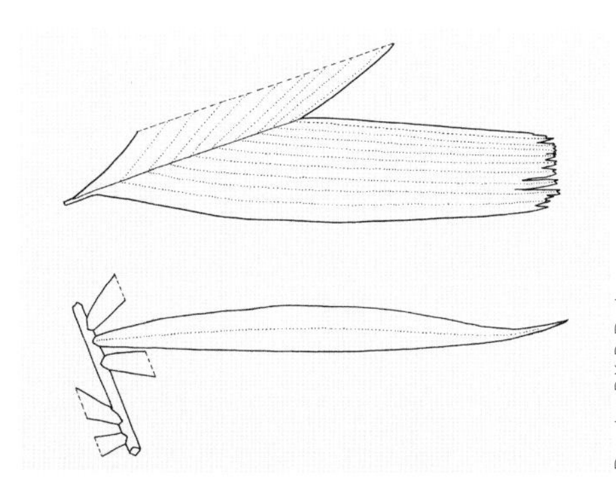

Drawing: P. K. R. Davies

Calyptrocalyx leaflet variation

Photo: W. J. Baker

Calyptrocalyx hollrungii colourful young leaf, near Finschhafen, PNG

Photo: W. J. Baker

Calyptrocalyx albertisianus inflorescence with young fruit, Wosimi River, Papua

Photo: W. J. Baker

Calyptrocalyx hollrungii inflorescence in fruit, near Finschhafen, PNG

Leaf pinnate, 20–290 cm long, 4–18 in crown, straight, leaf blade sometimes not divided into leaflets.

Sheath splitting to the base opposite the petiole, fibrous at the margins, 5–45 cm long, not forming crownshaft.

Petiole 5–40 cm long.

Leaflets (where blade divided) 2–36 each side of leaf rachis, 10–90 cm long, with lobed, toothed or pointed tips, arranged regularly or irregularly, horizontal.

Inflorescence between or below the leaves, unbranched (spicate), 15–200 cm long, some species producing more than one spike at a node.

Prophyll and peduncular bract similar, enclosing inflorescence in bud, not dropping off as inflorescence expands, peduncular bract projecting slightly from prophyll.

Peduncle longer than rachilla, 9–110 cm.

Rachillae usually slightly thicker than peduncle, straight.

Flowers in triads throughout the length of the rachilla, developing in pits, female flowers emerging from pits some time after the male flowers drop off.

Fruit red, pink or purple, globose to ellipsoid, 0.8–4 cm × 0.3–3 cm, stigmatic remains apical, flesh thin to quite thick, endocarp thin, closely adhering to seed.

Seed 1, globose to ellipsoid, endosperm ruminate or homogeneous.

Confused with:

- *Linospadix*: peduncular bract arising at top of peduncle and dropping off at maturity.
- *Sommieria*: leaves chalky white beneath, inflorescence branched, fruits with corky warts.

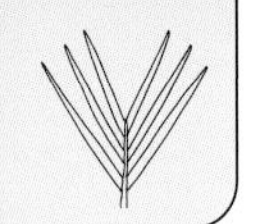

Calyptrocalyx – some examples

Calyptrocalyx albertisianus, Kikori, PNG

Calyptrocalyx albertisianus inflorescence in fruit, Lae Botanic Garden, PNG

Calyptrocalyx lauterbachianus leaf, near Finschhafen, PNG

Calyptrocalyx lauterbachianus inflorescence in fruit, near Finschhafen, PNG

Photo: J. Dransfield

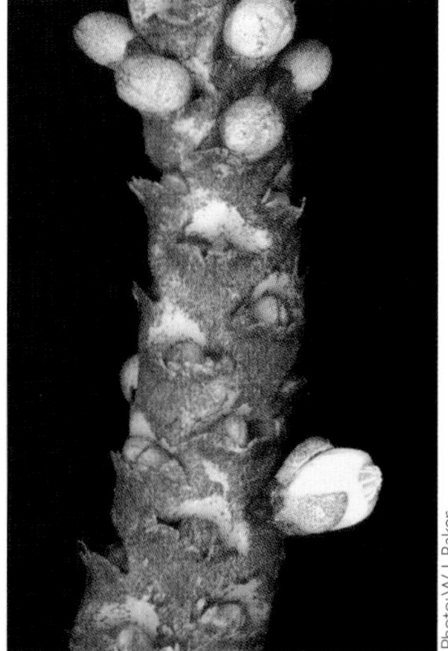

Photo: W. J. Baker

Clockwise, from top left

Calyptrocalyx micholitzii, near Timika, Papua

Calyptrocalyx doxanthus flowers, private collection, Bogor, Indonesia

Calyptrocalyx doxanthus, private collection, Bogor, Indonesia

Photo: J. Dransfield

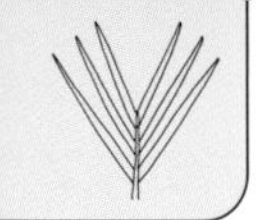

Clinostigma

Tall – pinnate – crownshaft – no spines – leaflets pointed

Look for:

- Robust, single-stemmed, canopy tree palm.
- Inflorescence below the leaves, resembling a horse's tail.
- Fruit asymmetric, with stigma to one side of apex.

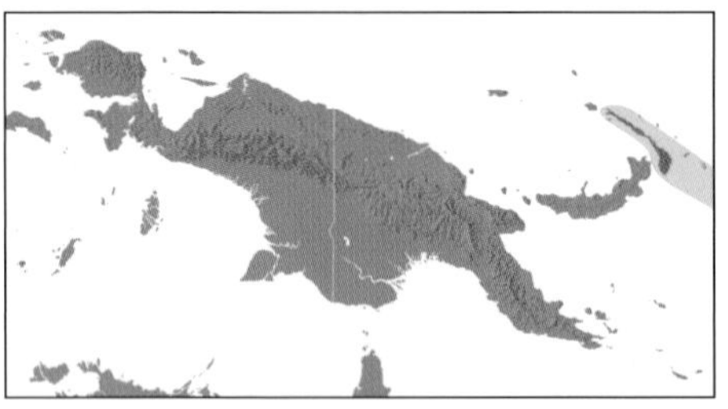

Distribution
Western Pacific, in New Guinea confined to New Ireland.

Habitat
Montane rainforest on limestone, c. 1300 m.

Number of species
Approximately 11, of which 1 in New Guinea.

Taxonomic accounts
Dransfield 1982.

Uses
None recorded.

Habit
Robust, single-stemmed tree palm, height c. 16 m, stem diameter c. 20 cm, slender crownshaft, monoecious.

Photo: J. Wood

Clinostigma collegarum, New Ireland, PNG

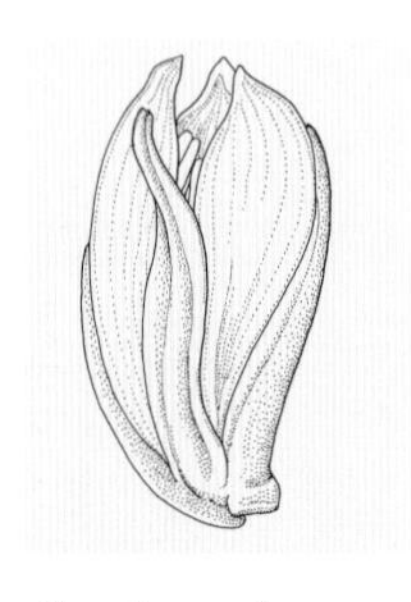

Drawing: L. T. Smith

Clinostigma collegarum male flower

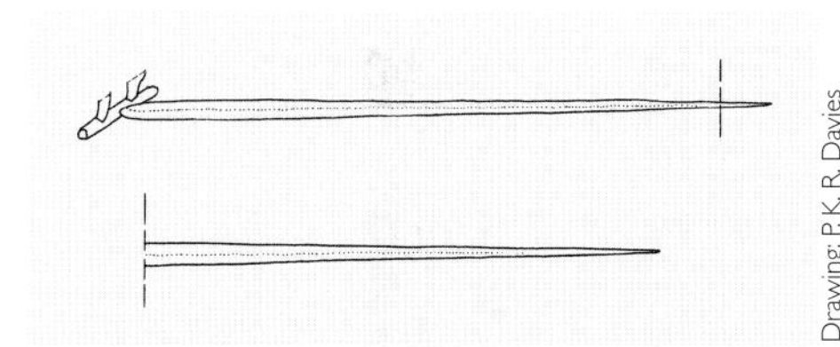
Drawing: P. K. R. Davies

Clinostigma leaflet

Photo: J. Wood

Clinostigma collegarum, New Ireland, PNG

Leaf pinnate, c. 400 m long, c. 13 in crown.

Sheath tubular, pale green, c. 200 cm long, forming crownshaft.

Petiole short, 60–70 cm long.

Leaflets c. 60 each side of leaf rachis, c. 120 cm long, with pointed tips, arranged regularly, ± pendulous.

Inflorescence below the leaves, branched to 3 orders, c. 70 cm long, resembling a horse's tail.

Prophyll and peduncular bract similar, enclosing inflorescence in bud, dropping off as inflorescence expands.

Peduncle shorter than inflorescence rachis, c. 12 cm, grossly swollen at base.

Rachillae slender and straight.

Flowers in triads throughout most of the length of the rachilla, pairs of male flowers towards the tip, not developing in pits, male flower asymmetrical, slightly larger than the female.

Fruit black, ellipsoid, c. 1.2 cm × c. 0.8 cm, stigmatic remains to one side of the apex, flesh juicy, endocarp thin, closely adhering to seed.

Seed 1, ellipsoid, endosperm homogeneous.

Photo: M. J. S. Sands

Clinostigma collegarum roots, New Ireland, PNG

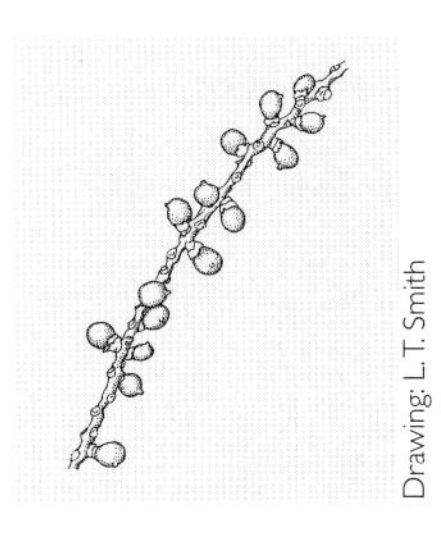
Drawing: L. T. Smith

Clinostigma collegarum fruit on rachilla

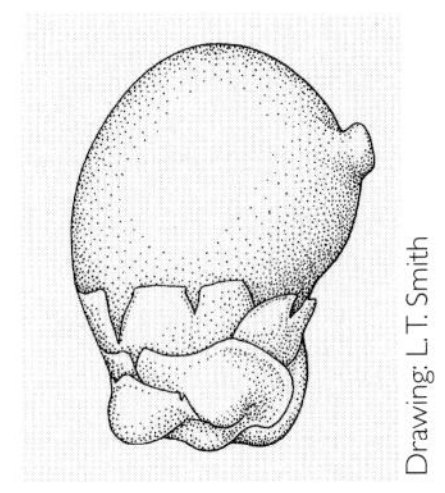
Drawing: L. T. Smith

Clinostigma collegarum fruit

Confused with:

- *Cyrtostachys*: inflorescence with widely spreading branches, flowers usually developing in pits.
- *Hydriastele*: leaflets with jagged tips, apical stigmatic remains.
- *Rhopaloblaste*: leaf with dark hairs on the leaf rachis, seedling leaf finely pinnate, inflorescence with widely spreading branches.

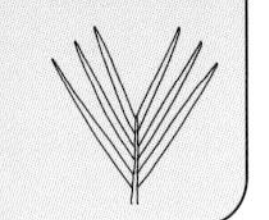

Cocos

Tall – pinnate – no crownshaft – no spines – leaflets pointed

Look for:

- Robust, single-stemmed tree palm with fibrous leaf sheaths; widely cultivated.
- Inflorescence between the leaves.
- Massive edible fruit with liquid contents.

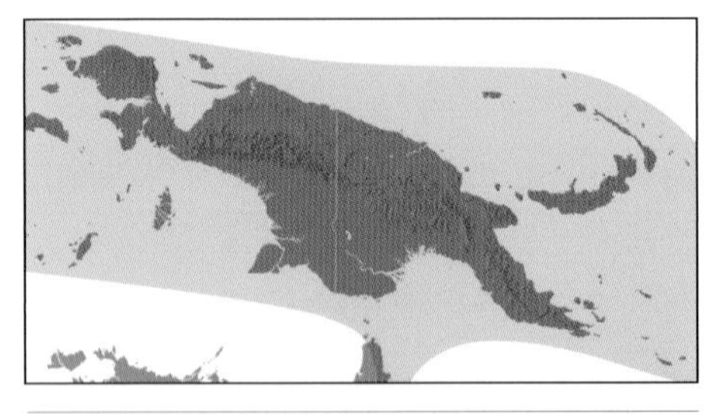

Distribution
Throughout the tropics.

Habitat
Coastal vegetation and cultivated areas.

Number of species
1 species.

Taxonomic accounts
Harries 1978.

Uses
Stem for construction; leaves for thatch, construction and weaving; shoot apex edible; seed edible; fruit fibre for weaving; endocarp used as containers.

Habit
Robust, single-stemmed tree palm, height to 30 m, stem diameter to 30 cm, crownshaft absent, monoecious.

Photo: W. J. Baker

Cocos nucifera, Madang, PNG

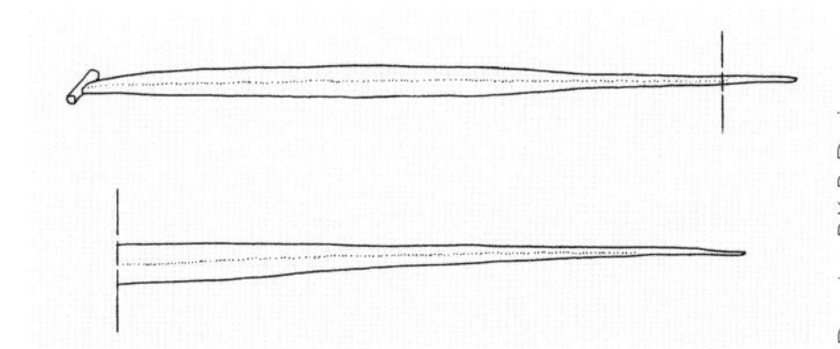

Drawing: P. K. R. Davies

Cocos leaflet

Photo: W. J. Baker

Cocos nucifera, Wosimi River, Papua

Photo: J. Dransfield

Cocos nucifera, Queensland, Australia

Leaf pinnate, to 600 cm long, up to 40 in crown, arching.

Sheath forming a fibrous network, c. 100 cm long, not forming crownshaft.

Petiole absent until sheath has disintegrated into fibres, then 100–130 cm long.

Leaflets 90–120 each side of leaf rachis, c. 100 cm long, with pointed tips, arranged regularly, often on sides.

Inflorescence between the leaves, branched to 3 orders, 100–150 cm long, branches widely spreading.

Prophyll short, hidden among leaf sheaths, peduncular bract much longer, woody, deeply grooved, enclosing inflorescence in bud and then splitting along its length, forming a cowl over the flowers, not dropping off as inflorescence expands.

Peduncle about the same length as inflorescence rachis, 50–75 cm.

Rachillae quite robust and ± straight.

Flowers in triads at the base of the rachillae, pairs of male flowers towards tip, not developing in pits, female flowers much larger than the male flowers.

Fruit yellow or green to brown, globose or ovoid, 20–30 cm × 10–30 cm, stigmatic remains apical, husk fibrous, endocarp thick, spherical, with three eyes.

Seed 1, globose, endosperm homogeneous, with a central hollow partially filled with fluid.

Confused with:

- *Orania*: leaflet tips jagged, fruit round, lacking fibrous husk.

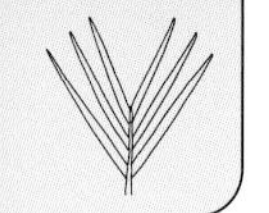

Cyrtostachys

Medium to tall – pinnate – crownshaft – no spines – leaflets pointed

Look for:

- Robust, single- or multi-stemmed tree palm, often with pendulous leaflets.
- Inflorescence below the leaves with widely spreading branches and flowers usually developing within pits.
- Fruit very small (≤1.2 cm), black, ellipsoid.

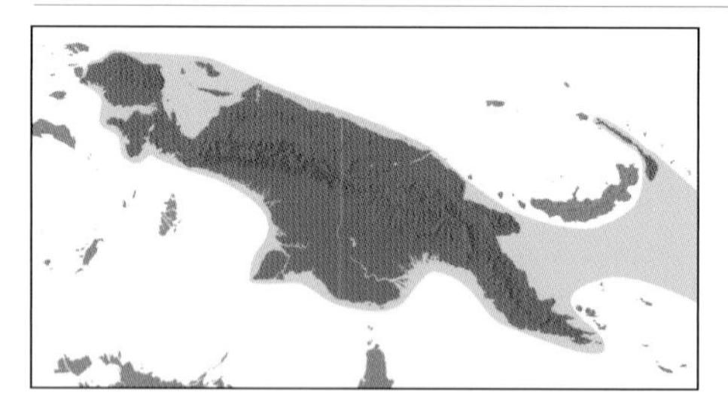

Photo: J. Dransfield

Cyrtostachys loriae, Timika, Papua

Distribution
Malaysia, Thailand, Sumatra and Borneo, New Guinea to Solomon Islands, widespread in New Guinea.

Habitat
Lowland to submontane rainforest, sea level to 1000 m.

Number of species
6, of which 5 in New Guinea.

Taxonomic accounts
Heatubun 2005.

Uses
Stem for floorboards.

Habit
Robust, single- or multi-stemmed tree palm, height to 25 m, stem diameter 5–30 cm, slender crownshaft, monoecious.

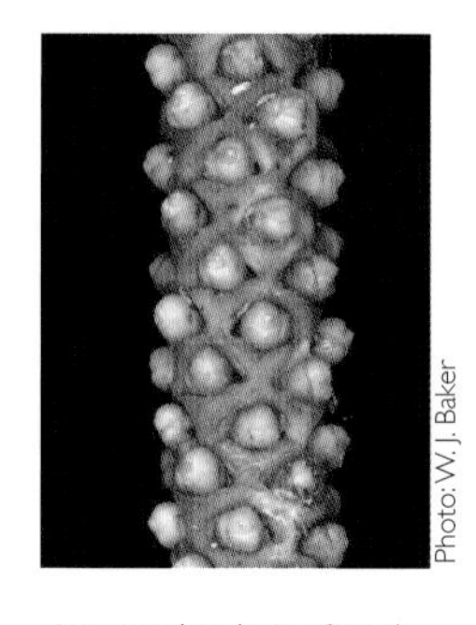

Photo: W. J. Baker

Cyrtostachys loriae female flowers, Kikori, PNG

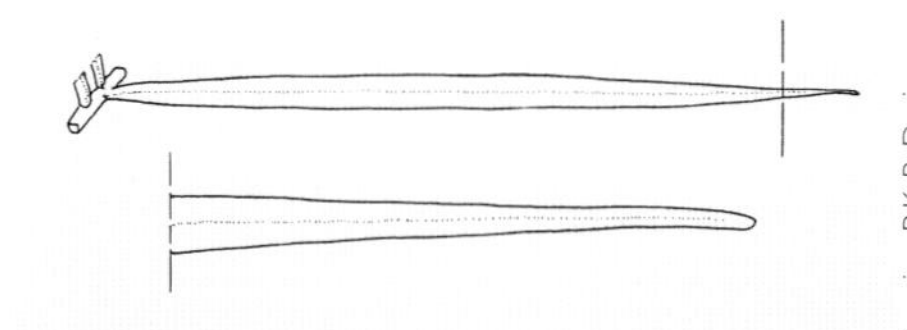

Drawing: P. K. R. Davies

Cyrtostachys leaflet

Photo: W. J. Baker

Cyrtostachys loriae inflorescence, Timika, Papua

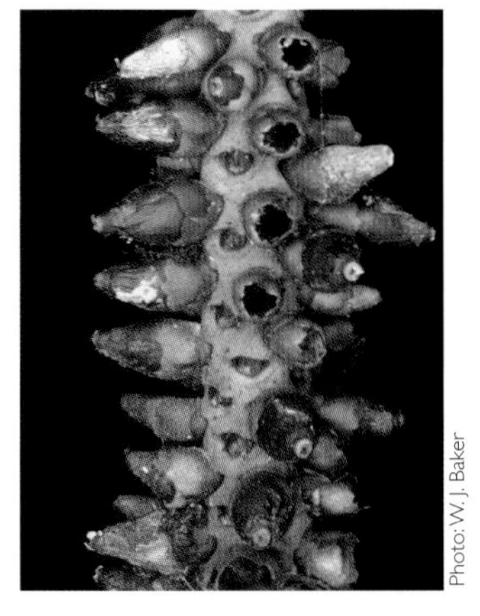

Photo: W. J. Baker

Cyrtostachys loriae fruit, Kikori, PNG

Leaf pinnate, 150–450 cm long, 6–12 in crown, straight or slightly arching.

Sheath tubular, pale green, to 200 cm long, forming crownshaft.

Petiole short, 10–25 cm long.

Leaflets 45–90 each side of leaf rachis, to 100 cm long, with pointed tips, arranged regularly, often pendulous.

Inflorescence below the leaves, branched to 3 orders, 60–90 cm long, branches widely spreading.

Prophyll and peduncular bract similar, enclosing inflorescence in bud, dropping off as inflorescence expands.

Peduncle shorter than inflorescence rachis, 10–20 cm, grossly swollen at base.

Rachillae slender to robust and straight.

Flowers in triads at the base of the branches, pairs of male flowers towards tip, usually developing in pits.

Fruit black, narrow ellipsoid, 0.8–1.2 cm × 0.4–0.5 cm, stigmatic remains apical, flesh thin and juicy, endocarp thin, closely adhering to seed.

Seed 1, ellipsoid, endosperm homogeneous.

Confused with:

- *Clinostigma*: inflorescence resembling a horse's tail, flowers not developing in pits, fruit asymmetric, with stigma to one side of apex.
- *Hydriastele*: jagged leaflet tips, inflorescence resembling a horse's tail, flowers not developing in pits.
- *Rhopaloblaste*: leaf with dark hairs on the leaf rachis, seedling leaf finely pinnate, flowers not developing in pits, fruit larger, endosperm deeply ruminate.

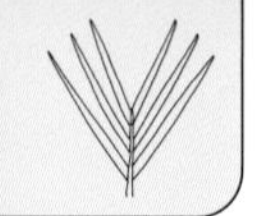

Dransfieldia

Small – pinnate – crownshaft – no spines – leaflets pointed

Look for:

- Slender, multi-stemmed (rarely single-stemmed), mid-storey tree palm.
- Inflorescence below the leaves, prophyll and sometimes peduncular bract not dropping off.
- Fruit black when ripe.

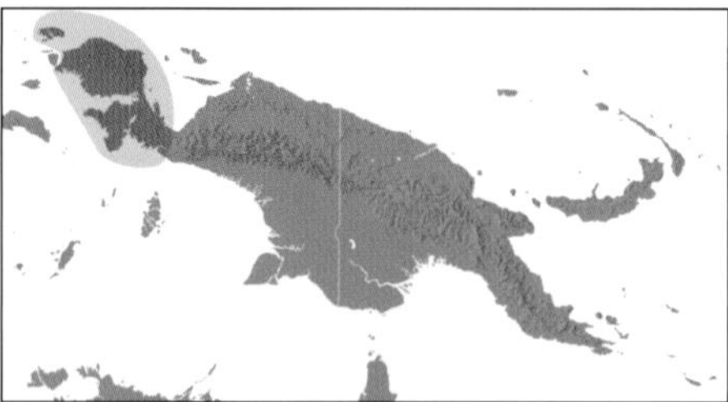

Distribution
Endemic to New Guinea, restricted to the far west.

Habitat
Lowland rainforest, sea level to 180 m.

Number of species
1 species.

Taxonomic accounts
Baker et al. 2006.

Uses
Stems for harpoons; leaves for thatch.

Habit
Slender, multi-stemmed (rarely single-stemmed) tree palm, height to 10 m, stem diameter 2–5 cm, crownshaft present, monoecious.

Photo: W. J. Baker

Dransfieldia micrantha, Wosimi River, Papua

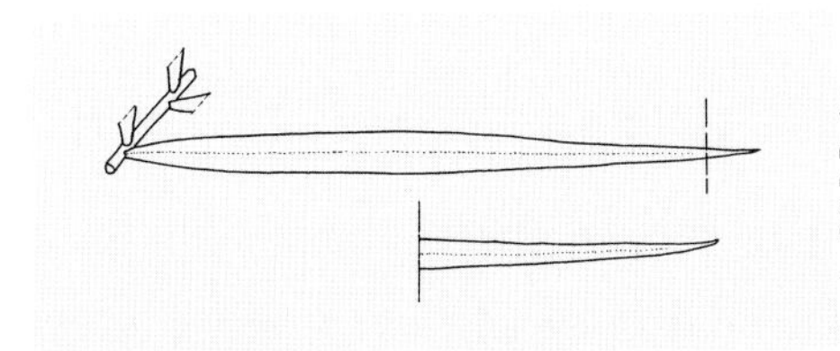

Drawing: P. K. R. Davies

Dransfieldia leaflet

Photo: W. J. Baker

Dransfieldia micrantha, Wosimi River, Papua

Photo: W. J. Baker

Dransfieldia micrantha inflorescences, Wosimi River, Papua

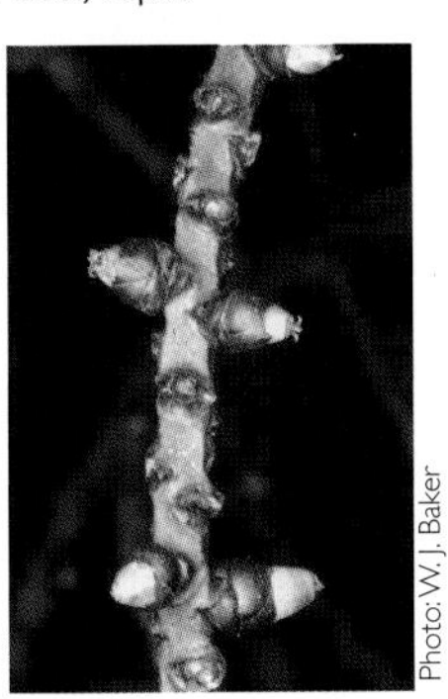

Photo: W. J. Baker

Dransfieldia micrantha recently pollinated female flowers, Wosimi River, Papua

Leaf pinnate, 100–200 cm long, 4–7 in crown, ± straight.

Sheath tubular, green with dark indumentum, 30–45 cm long, forming narrow crownshaft.

Petiole 10–20 cm long.

Leaflets 12–30 each side of leaf rachis, to 80 cm long, with pointed tips, arranged regularly, horizontal.

Inflorescence below the leaves, branched to 2 (rarely 3) orders, 34–60 cm long, branches spreading.

Prophyll and peduncular bract similar, but peduncular bract projecting from prophyll and enclosing inflorescence in late bud, prophyll and sometimes peduncular bract not dropping off as inflorescence expands.

Peduncle same length as or slightly longer than inflorescence rachis, 12–26 cm.

Rachillae quite robust and curving.

Flowers in triads throughout the length of the rachillae, not developing in pits, male flowers bullet-shaped.

Fruit black, ellipsoid, 1.5–1.6 cm × 0.7–1 cm, stigmatic remains apical, flesh thin, endocarp thin, closely adhering to seed.

Seed 1, ovoid with flattened base, endosperm ruminate.

Confused with:

- *Drymophloeus*: leaflet tips jagged, stilt roots present (except in *D. hentyi*).
- *Heterospathe*: no crownshaft, leaf sheaths fibrous.
- *Ptychosperma*: leaflet tips jagged, primary inflorescence bracts dropping off.
- *Rhopaloblaste*: tall, single-stemmed tree palm, leaf with dark hairs on the leaf rachis, seedling leaf finely pinnate, inflorescence with short peduncle, primary inflorescence bracts dropping off.

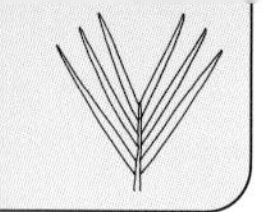

Drymophloeus

Small to medium – pinnate – crownshaft – no spines – leaflets jagged

Look for:

- Slender, usually single-stemmed, under- to mid-storey tree palm, usually with stilt roots.
- Inflorescence below the leaves, prophyll usually not dropping off.
- Fruit small (10–24 mm), red when ripe.

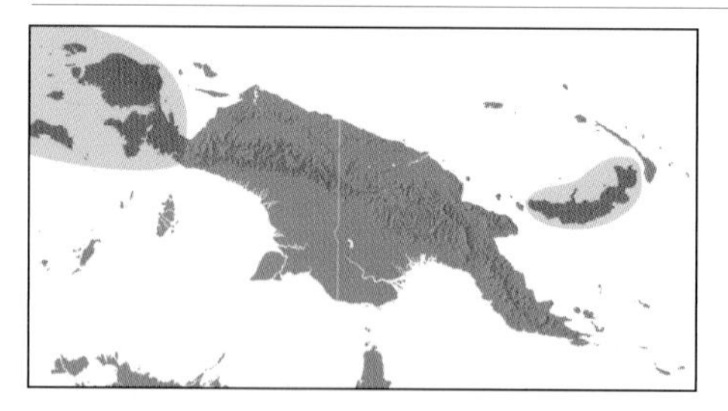

Distribution
Moluccas to Solomon Islands, in New Guinea restricted to the far west and New Britain.

Habitat
Lowland rainforest, sea level to 1200 m.

Number of species
Approximately 7, of which 3 in New Guinea.

Taxonomic accounts
Zona 1999.

Uses
Stems for spears and arrow tips; leaves for wrapping sago.

Habit
Slender, usually single-stemmed tree palm, usually with a cone of stilt roots (except in *D. hentyi*), height to 10 m, stem diameter 1–8 cm, crownshaft present, monoecious.

Drymophloeus oliviformis, Wosimi River, Papua

Drymophloeus oliviformis inflorescence in fruit, Wosimi River, Papua

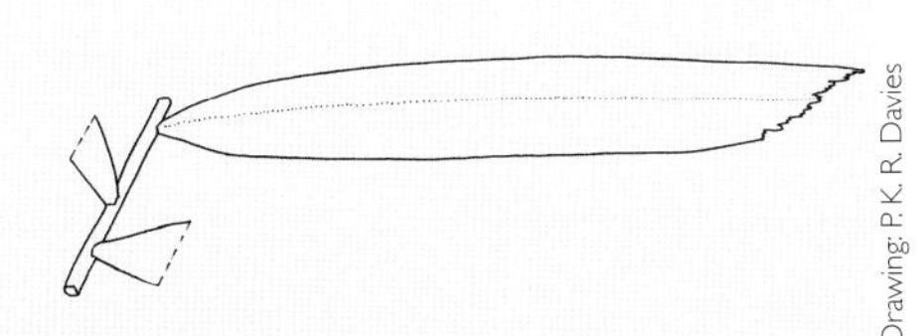

Drawing: P. K. R. Davies

Drymphloeus leaflet

Photo: J. Dransfield

Drymophloeus hentyi, Hoomaluhia Botanical Garden, Hawaii

Photo: W. J. Baker

Drymophloeus oliviformis stilt roots, Wosimi River, Papua

Photo: W. J. Baker

Drymophloeus litigiosus male flowers, Wandammen Peninsula, Papua

Leaf pinnate, 86–300 cm long, 6–13 in crown, straight to arching.

Sheath tubular, 15–80 cm long, forming crownshaft.

Petiole 15–61 cm long.

Leaflets 7–21 each side of leaf rachis, to 71 cm long, narrowly to broadly wedge-shaped, with jagged tips, arranged regularly, horizontal.

Inflorescence below the leaves, branched 1–3 orders, 18–75 cm long, branches spreading.

Peduncular bract longer than and projecting from prophyll, enclosing inflorescence in bud, prophyll (and sometimes peduncular bract) usually not dropping off as inflorescence expands.

Peduncle sometimes longer than inflorescence rachis, 7–38 cm.

Rachillae slender to quite robust, straight to curving.

Flowers in triads throughout the length of the rachillae, not developing in pits, male flowers larger than female flowers, bullet-shaped.

Fruit red, ovoid, 1–2.4 cm × 0.5–1.3 cm, stigmatic remains apical, flesh quite thick, juicy, filled with irritant needle crystals, endocarp pale, thin, smooth or 5–lobed.

Seed 1, conforming to endocarp, endosperm homogeneous or ruminate.

Confused with:

- *Brassiophoenix*: broad wedge-shaped leaflets, no stilt roots, orange, medium-sized fruit (3.5 cm long) endocarp pale with conspicuous grooves and ridges.
- *Dransfieldia*: no stilt roots, leaflet tips pointed, fruit black.
- *Ptychococcus*: moderate to robust palm, no stilt roots, fruit quite large (4–6 cm), endocarp black or dark brown with conspicuous grooves and ridges.
- *Ptychosperma*: no stilt roots, primary inflorescence bracts dropping off, endocarp frequently grooved.

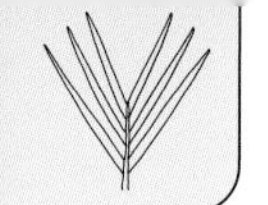

Heterospathe

Small to medium – pinnate – no crownshaft – no spines – leaflets pointed

Look for:

- Varied under- and mid-storey palms, some stemless, usually with fibrous leaf sheaths.
- Inflorescence between or below the leaves, branched 1 to 3 orders.
- Peduncular bract much longer than prophyll.

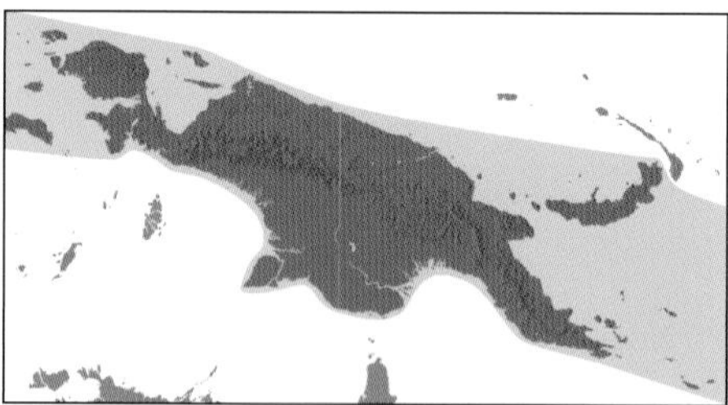

Distribution
Philippines to Moluccas, through Papuasia to Fiji, widespread in New Guinea.

Habitat
Mainly in submontane and montane rainforest, less frequently in lowlands, from sea level to 2300 m.

Number of species
Approximately 40, of which 19 in New Guinea.

Taxonomic accounts
Moore 1969, checklist in Govaerts & Dransfield 2005.

Uses
Stem for bows and spears; shoot apex edible; seed as a substitute for betel.

Habit
Moderate to quite robust single- and multi-stemmed tree palm, height to 10 m, some species stemless, stem diameter 3–7.5 cm, no crownshaft, monoecious.

Photo: W. J. Baker

Heterospathe sp., Southern Highlands, PNG

Photo: W. J. Baker

Heterospathe sp. inflorescence, Southern Highlands, PNG

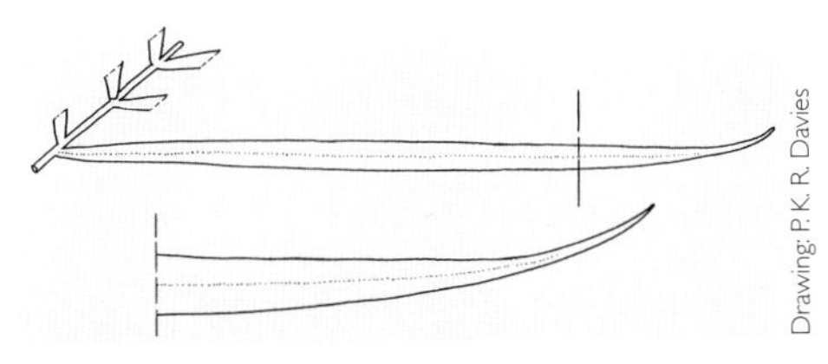

Heterospathe leaflet

Heterospathe sp. crown, Southern Highlands, PNG

Heterospathe sp. inflorescence, near Finschhafen, PNG

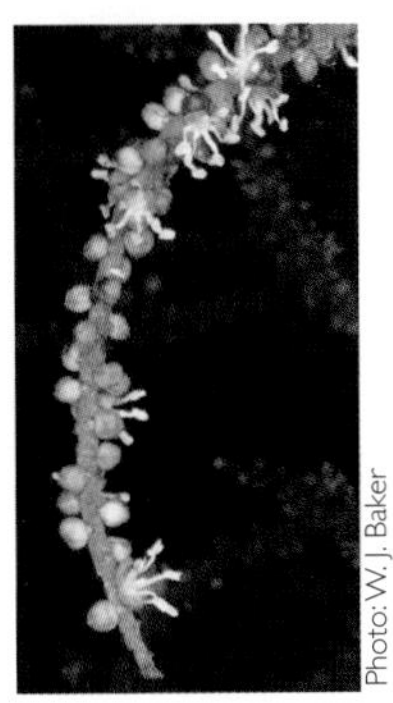

Heterospathe sp. rachilla, Southern Highlands, PNG

Leaf pinnate, 90–330 cm long, 7–18 in crown, ± straight.

Sheath splitting to the base opposite the petiole, fibrous at the margins, 20–55 cm long, not forming crownshaft.

Petiole 20–150 cm long.

Leaflets 9–56 each side of leaf rachis, to 80 cm long, with pointed tips, arranged regularly (rarely somewhat irregularly), horizontal.

Inflorescence between or below the leaves, branched 1–3 (possibly 4) orders, 46–130 cm long, branches widely spreading.

Peduncular bract larger than and projecting from the prophyll, peduncular bract alone enclosing inflorescence in late bud, splitting, but usually remaining attached as inflorescence expands.

Peduncle as long as or much longer than the inflorescence rachis, 30–110 cm.

Rachillae slender and straight or curving.

Flowers in triads throughout the length of the rachilla, not developing in pits.

Fruit orange to red, globose to ellipsoid, 1–2.2 cm × 0.5–1.2 cm, stigmatic remains apical to lateral, flesh thin, endocarp thin, closely adhering to seed.

Seed 1, ellipsoid to globose, endosperm ruminate.

Confused with:

- *Dransfieldia*: crownshaft present.
- *Rhopaloblaste*: leaf with dark hairs on the leaf rachis, crownshaft present, seedling leaf finely pinnate, inflorescence with short peduncle, primary inflorescence bracts dropping off.

Heterospathe – some examples

Heterospathe macgregorii grows only on river banks, Kikori River, PNG

Heterospathe macgregorii fruit, Kikori River, PNG

Heterospathe sp., near Tabubil, PNG

Heterospathe sp. inflorescence in fruit, near Tabubil, PNG

Heterospathe humilis, near Finschhafen, PNG

Heterospathe humilis inflorescence, near Finschhafen, PNG

Heterospathe delicatula, Lyon Arboretum, Hawaii

Heterospathe delicatula inflorescence, Lyon Arboretum, Hawaii

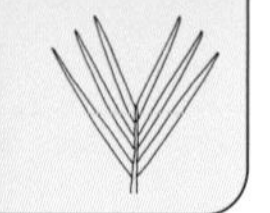

Hydriastele

Very small to tall – pinnate – crownshaft – no spines – leaflets jagged

Look for:

- Varied understorey to canopy tree palms, leaflets tips jagged (except in *H. costata*).
- Inflorescence below the leaves, resembling a horse's tail, unless unbranched (spicate).
- Fruit small, red or black (0.5–1.5 cm).

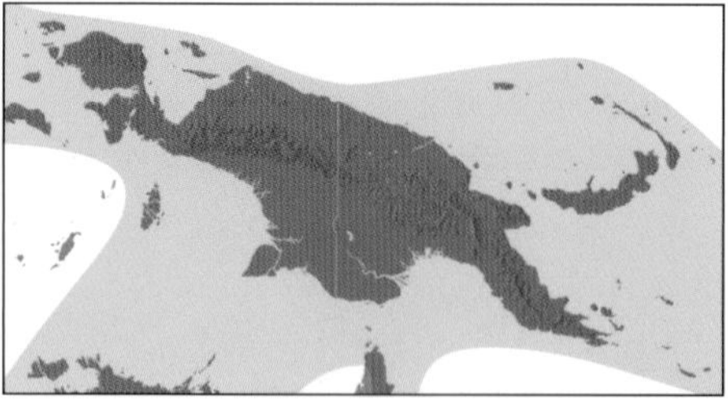

Distribution
Sulawesi to Fiji and Australia, widespread in New Guinea.

Habitat
Lowland to submontane rainforest, sea level to 2200 m.

Number of species
Approximately 48, of which 32 in New Guinea.

Taxonomic accounts
Baker & Loo 2004, Loo et al. 2006.

Uses
Stem for floorboards, roofing, beds, spears, bows and arrows; inflorescences for brooms; seed as betel substitute.

Habit
Very slender to robust, single- or multi-stemmed tree palms, height to 25 m, stem diameter 0.7–35 cm, swollen in some species, crownshaft present, monoecious.

Photo: W. J. Baker

Hydriastele chaunostachys, near Finschhafen, PNG

Photo: W. J. Baker

Hydriastele sp. fruit, Kikori, PNG

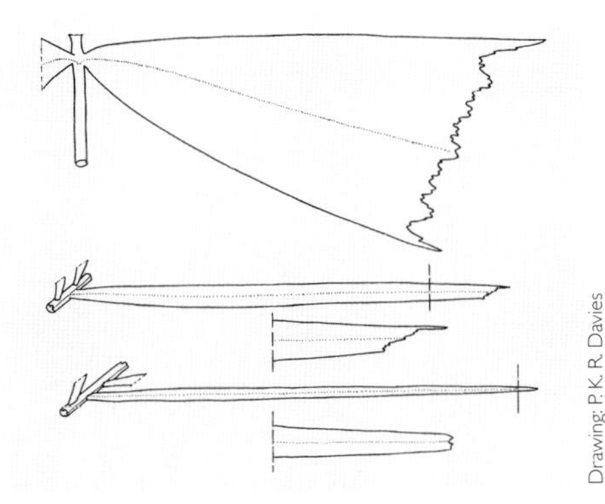

Drawing: P. K. R. Davies

Hydriastele leaflet variation

Photo: W. J. Baker

Hydriastele costata crown, Lae Botanic Garden, PNG

Leaf pinnate, 38–450 cm long, 4–20 in crown, strongly arching in many species.

Sheath tubular, 15–200 cm long, forming crownshaft, often with conspicuous indumentum of various kinds.

Petiole 15–80 cm long, leaf blade occasionally not divided into leaflets (also in some juveniles).

Leaflets (where blade divided) 2–74 each side of leaf rachis, 10–120 cm long, linear to wedge-shaped, with jagged tips (tips rarely pointed, e.g. *H. costata*), arranged regularly or irregularly, ascending, horizontal or pendulous.

Inflorescence below the leaves, spicate or branched 1–3 orders, 9–100 cm long, branches swept forward and resembling a horse's tail or brush.

Prophyll and peduncular bract similar, enclosing inflorescence in bud, dropping off as inflorescence expands.

Peduncle shorter than inflorescence rachis, 2–15 cm.

Rachillae slender and straight or arching.

Flowers in triads throughout the length of the rachilla, not developing in pits, male flowers larger than female flowers, asymmetrical and more conspicuous, with pink or white pointed petals.

Fruit red or black, narrowly to broadly ellipsoid, 0.5–1.5 cm × 0.5–1 cm, stigmatic remains apical, flesh thin, endocarp thin, closely adhering to seed.

Seed 1, ellipsoid, endosperm homogeneous or ruminate.

Confused with:

- *Clinostigma*: leaflet tips pointed, fruit asymmetric, with stigma to one side of apex.
- *Cyrtostachys*: leaflet tips pointed, inflorescence with widely spreading branches, flowers usually developing in pits.
- *Drymophloeus*: stilt roots, inflorescence with spreading branches.
- *Ptychosperma*: stilt roots, inflorescence with spreading branches.
- *Rhopaloblaste*: leaf with dark hairs on the leaf rachis, leaflet tips pointed, seedling leaf finely pinnate, inflorescence with widely spreading branches.

Hydriastele – some examples

Hydriastele chaunostachys inflorescence, near Finschhafen, PNG

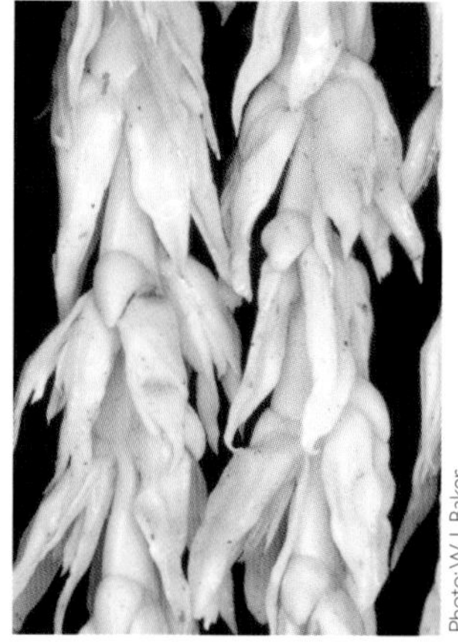

Hydriastele sp. triads with open male flowers, Kikori, PNG

Hydriastele chaunostachys female flowers, near Finschhafen, PNG

Hydriastele microspadix, Kikori, PNG

Hydriastele costata female flowers, Kikori, PNG

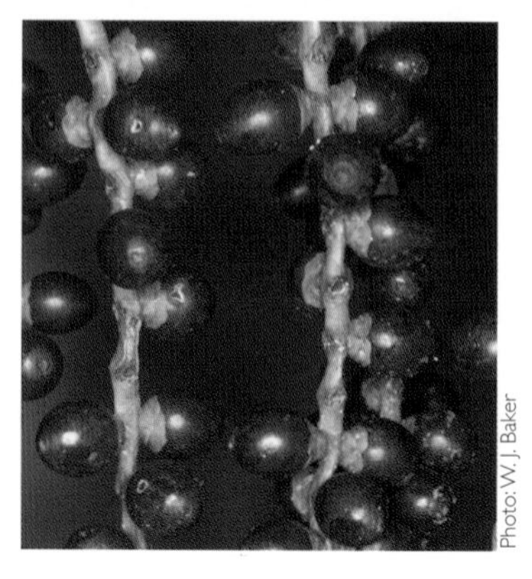

Hydriastele microspadix fruit, Nadzab, PNG

Hydriastele dransfieldii, Biak, Papua

Hydriastele dransfieldii fruit, private collection, Bogor, Indonesia

Hydriastele rhopalocarpa, Timika, Papua

Hydriastele costata fruit, Wandammen Peninsula, Papua

Hydriastele costata, Mubi River, PNG

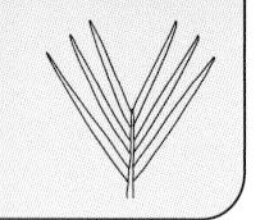

Linospadix

Very small – pinnate – no crownshaft – no spines – leaflets slightly toothed or pointed

Look for:

- Very slender, single- or multi-stemmed understorey palm, leaf blade sometimes forked.
- Inflorescences between the leaves, unbranched (spicate), peduncular bract arising at top of peduncle and dropping off at maturity.
- Flowers developing within pits in the inflorescence spike.

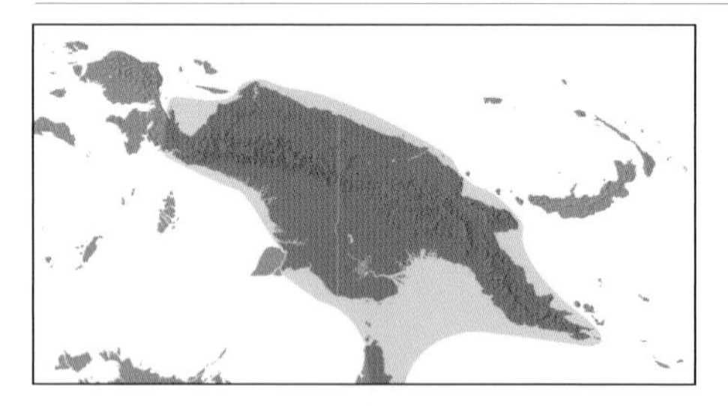

Distribution
New Guinea and Australia, widespread in New Guinea.

Habitat
Lowland to montane rainforest, sea level to 1350 m.

Number of species
9, of which 2 in New Guinea.

Taxonomic accounts
Dowe & Ferrero 2001.

Uses
Stem for stirring sago; leaves for wrapping food.

Habit
Very slender, single- or multi-stemmed palms, height to 2 m, stem diameter 0.4–10 cm, no crownshaft, monoecious.

Linospadix albertisianus, Mt. Bosavi, PNG

Linospadix albertisianus fruit, Mt. Bosavi, PNG

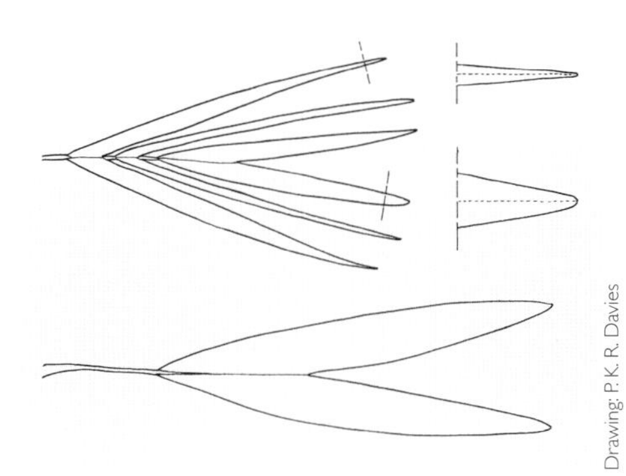

Drawing: P. K. R. Davies

Linospadix leaf variation

Photo: W. J. Baker

Linospadix albertisianus inflorescence with peduncular bract beginning to split, Mt. Bosavi, PNG

Leaf pinnate, 20–90 cm long, 4–12 in crown, ± straight.

Sheath splitting to the base opposite the petiole, 6–12 cm long, not forming crownshaft.

Petiole 5–12 cm long.

Leaf blade sometimes not divided into leaflets and appearing forked.

Leaflets (where blade divided) up to 3 each side of leaf rachis, 10–30 cm long, with slightly pointed or toothed tips, arranged regularly or irregularly, horizontal.

Inflorescence between the leaves, unbranched (spicate), 20–65 cm long.

Peduncular bract similar to prophyll, but attached at the base of the tip of the peduncle and projecting far from the prophyll, splitting as inflorescence matures and dropping off, a scar remaining.

Peduncle longer than rachilla, 10–40 cm.

Rachilla slender, straight.

Flowers in triads throughout the length of the rachilla, developing in pits, female flowers emerging from pits some time after the male flowers drop off.

Fruit red, narrow ellipsoid, 1.5–3 cm × 0.3–0.6 cm, stigmatic remains apical, flesh thin and juicy, endocarp thin, closely adhering to seed.

Seed 1, narrow ellipsoid, endosperm homogeneous.

Photo: W. J. Baker

Linospadix albertisianus inflorescence with male flowers, Mt. Bosavi, PNG

Confused with:

- *Calyptrocalyx*: peduncular bract arising near base of peduncle.

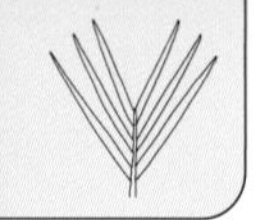

Metroxylon

Tall – pinnate – no crownshaft – spiny – leaflets pointed

Look for:

- Robust, single- or multi-stemmed tree palm of swamp forest, usually with spiny leaves.
- Inflorescences above the leaves.
- Stem dying after flowering.

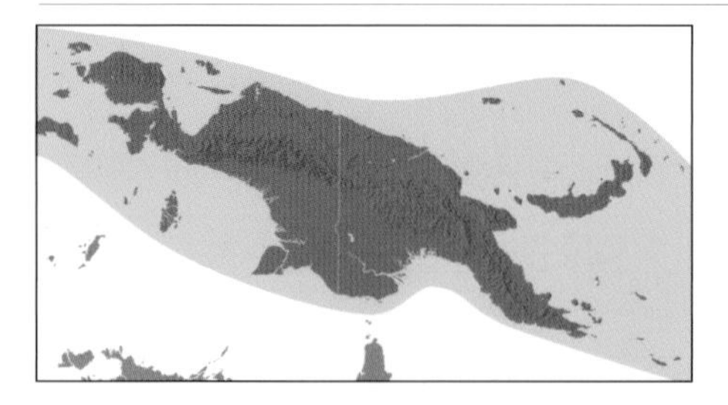

Distribution
New Guinea and western Pacific, widespread in New Guinea, widely cultivated in South-East Asia.

Habitat
Lowland swamp forest at sea level, sometimes cultivated on swampy ground at higher elevations.

Number of species
7, of which 2 in New Guinea.

Taxonomic accounts
Rauwerdink 1986, Kjær et al. 2004.

Uses
Stem for the production of a staple source of starch, construction and for breeding sago grubs; shoot apex edible; leaves for thatch; petioles for the walls of houses; sheaths for sago processing and sometimes painted for decoration; spines for body piercing; fruit for decoration.

Habit
Robust, single- or multi-stemmed tree palm, height to 20 m, stem diameter 15–60 cm, no crownshaft, stems dying after flowering, hermaphrodite.

Photo: W. J. Baker

Metroxylon sagu, near Lae, PNG

Photo: W. J. Baker

Metroxylon sagu fruit, Wosimi River, Papua

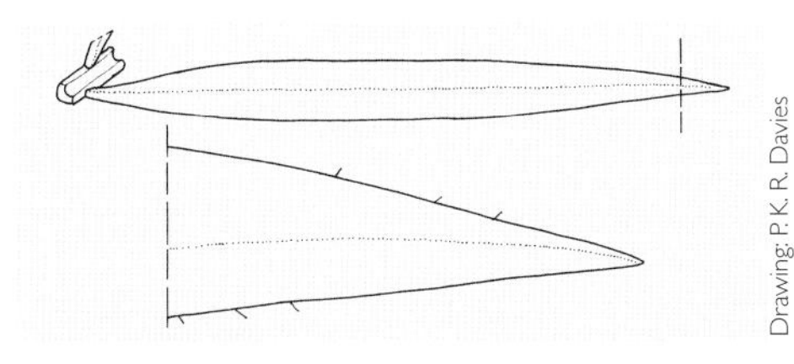

Drawing: P. K. R. Davies

Metroxylon leaflet

Leaf pinnate, to c. 800 cm long, 8–12 in crown, armed with spines and bristles, erect.

Sheath splitting to the base opposite the petiole, massive, not forming crownshaft.

Petiole c. 200 cm long.

Leaflets c. 80 each side of leaf rachis, to 150 cm long, with pointed tips, arranged regularly, ± horizontal.

Photo: W.J. Baker

Metroxylon sagu dead stems after flowering, Wosimi River, Papua

Photo: W.J. Baker

Metroxylon sagu flowers, Timika, Papua

Inflorescence above the leaves, many inflorescences produced simultaneously, individual inflorescences branched to 2 orders, c. 400 cm long, branches widely spreading, flowering resulting in the death of the stem.

Primary bracts similar, not enclosing inflorescence in late bud, not dropping off as inflorescence expands.

Peduncle shorter than inflorescence rachis, up to 150 cm.

Rachillae robust, sausage-like, straight, with conspicuous bracts.

Flowers in pairs throughout the length of the rachilla, developing in pits formed by rachilla bracts.

Fruit brown, globose, 2.5–3.5 cm × 3–4 cm, with vertical rows of shiny scales, stigmatic remains apical, flesh rather spongy and dry.

Seed 1, enclosed in a fleshy coat (sarcotesta), globose, with sarcotesta intruding into one side of the seed, endosperm homogeneous.

Photo: W.J. Baker

Metroxylon sagu leaf sheaths, Wosimi River, Papua

Photo: W.J. Baker

Metroxylon sagu inflorescence, Timika, Papua

Confused with:

- *Nypa*: not spiny, never forming erect stem, palm of mangrove swamp.
- *Pigafetta*: single-stemmed canopy tree, fruit very small (0.8–1 cm), stem not dying after flowering.

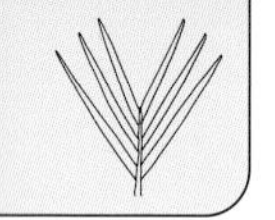

Nypa

Stemless – pinnate – no crownshaft – no spines – leaflets pointed

Look for:

- Robust palm with horizontal, forking stem, abundant in mangrove.
- Inflorescence between the leaves, all bracts orange and rubbery.
- Fruiting head club-like, breaking into large brown fruits (8–10 cm long).

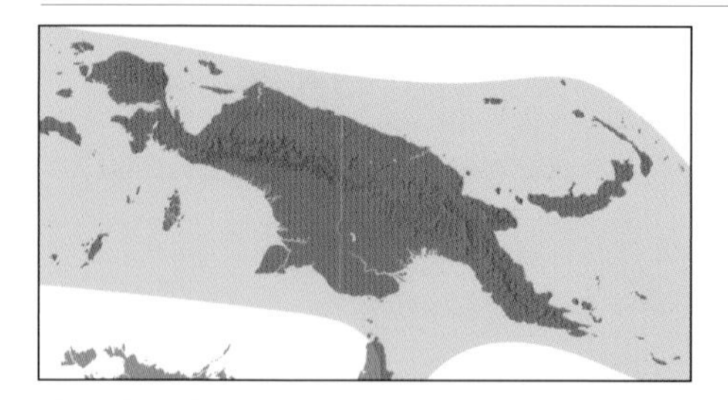

Distribution
South China to India, through South-East Asia to Australia and Solomon Islands, widespread in New Guinea.

Habitat
Mangrove forest.

Number of species
1 species.

Taxonomic accounts
Checklist in Govaerts & Dransfield 2005.

Uses
Leaves for thatch; leaf epidermis for cigarette papers.

Habit
Robust palm with horizontal, forking stem, abundant in mangrove, monoecious.

Photo: S. Barrow

Nypa fruticans, Wosimi River, Papua

Photo: J. Dransfield

Nypa fruticans fruit, Brunei

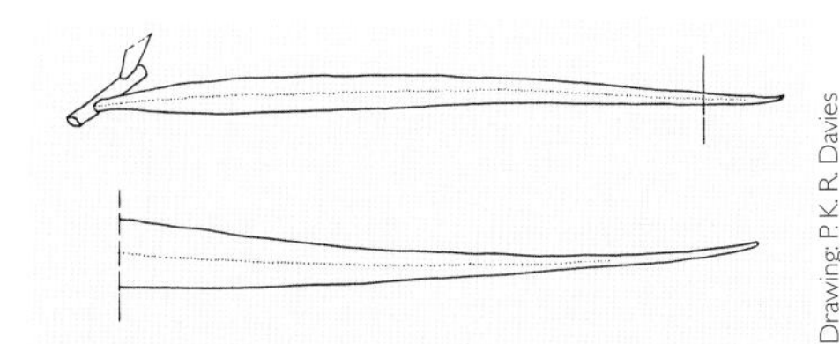
Drawing: P. K. R. Davies

Nypa leaflet

Leaf pinnate, to 800 cm long, 10–12 in crown, erect.

Sheath open at base, to 200 cm long, not forming crownshaft.

Petiole long, to 400 cm long.

Leaflets to 60 each side of leaf rachis, to 130 cm long, with pointed tips, arranged regularly, horizontal, forked scales abundant on lower surface of leaflet midrib.

Inflorescence between the leaves, branched to 5 orders, to 150 cm long.

Primary bracts and all subsequent bracts similar, orange and rubbery, not dropping off as inflorescence expands.

Peduncle shorter than inflorescence rachis, c. 30 cm.

Inflorescence ending in a round head of female flowers, male rachillae lateral, sausage-like, densely covered in flowers.

Flowers male flowers very small, bright yellow, female flowers larger, brown.

Fruit dark brown, in a large head at the end of the peduncle, individual fruits wedge-like, 8–10 cm × 4–7 cm, deeply grooved, husk fibrous, endocarp hard and thick.

Seed 1, endosperm homogeneous or ruminate.

Photo: J. Dransfield

Nypa fruticans inflorescence, Malaysia

Photo: J. Dransfield

Nypa fruticans male flowers, Malaysia

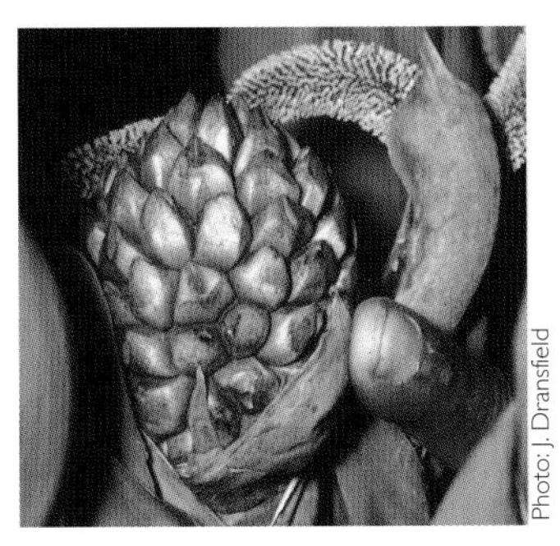
Photo: J. Dransfield

Nypa fruticans female flowers, Malaysia

Confused with:

- *Metroxylon*: tree palm, spiny, not in mangrove, stem dying after flowering.

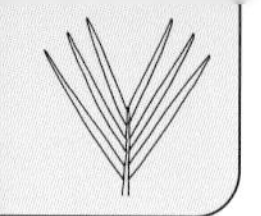

Orania

Medium to tall – pinnate – no crownshaft – no spines – leaflets jagged, pale beneath

Look for:

- Usually robust single-stemmed tree palms of mid-storey to canopy with fibrous leaf sheaths.
- Inflorescence between the leaves, peduncular bract much longer than prophyll.
- Fruit round, quite large (3.5–8 cm), occasionally bilobed or trilobed.

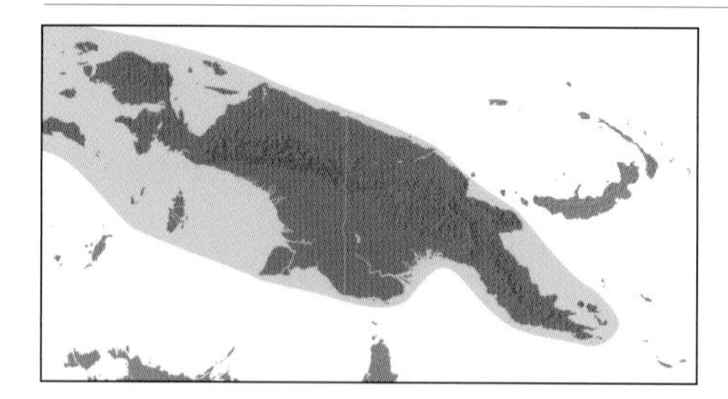

Distribution
Madagascar, Thailand to New Guinea, widespread in New Guinea.

Habitat
Lowland to montane rainforest, sea level to 1800 m.

Number of species
27, of which 21 in New Guinea.

Taxonomic accounts
Essig 1980, Keim 2003.

Uses
Stem for construction, bows and arrows; shoot apex and fruit reported to be poisonous.

Habit
Usually robust, single-stemmed tree palms, height to 40 m, stem diameter 10–40 cm, monoecious.

Photo: W. J. Baker

Orania lauterbachiana, Kikori, PNG

Photo: W. J. Baker

Orania regalis germinating seed, Wandammen Peninsula, Papua

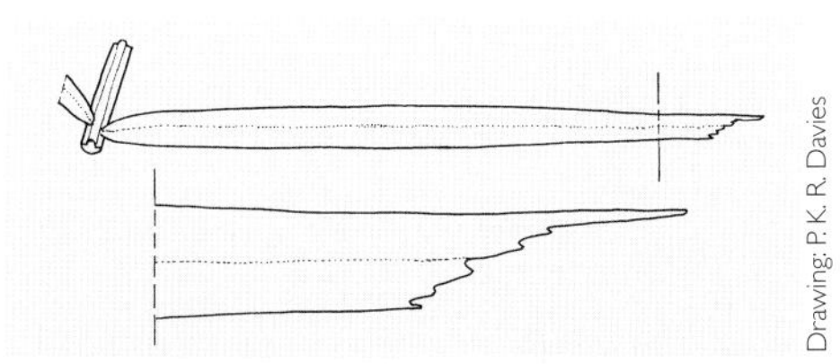

Orania leaflet

Orania lauterbachiana inflorescence, Kikori, PNG

Orania palindan fruit, Wandammen Peninsula, Papua

Orania sp. male flowers, Southern Highlands, PNG

Orania sp. female flowers, Southern Highlands, PNG

Leaf pinnate, usually spirally arranged or arranged in a fan (distichous), 200–700 cm long, 6–20 in crown, arching or horizontal.

Sheath to 75 cm long, splitting to base opposite the petiole, not forming a crownshaft.

Petiole 20–150 cm long.

Leaflets 20–80 each side of leaf rachis, to 150 cm long, with jagged tips, arranged regularly or irregularly, sometimes pendulous, covered with whitish indumentum on the undersurface.

Inflorescence between the leaves, branched 1–4 orders, to 250 cm long.

Prophyll hidden among leaf sheaths, peduncular bract much large than prophyll, enclosing the inflorescence in bud, strongly beaked, not dropping off as inflorescence expands, but often decaying later.

Peduncle longer than inflorescence rachis, 75–150 cm.

Rachillae generally slender, straight, curved or zig-zag.

Flowers in triads at the base of the branches, pairs of male flowers towards tip, not developing in pits.

Fruit green, yellow or orange-red, large, round, bilobed or trilobed (depending on number of seeds), single-seeded fruit (or individual lobes) 3.5–8 cm × 3.5–8 cm, stigmatic remains usually basal, flesh dry, endocarp smooth with a basal heart-shaped button.

Seed 1 (rarely 2–3), globose, endosperm homogeneous.

Confused with:

- *Arenga*: usually multi-stemmed, leaflet bases asymmetrical and V-shaped in section, stems dying after flowering, fruit rather small.
- *Cocos*: leaflets pointed, same colour on both surfaces, fruit very large.

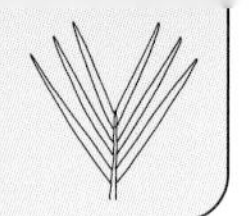

Orania – some examples

Photo: W. J. Baker

Orania palindan, Wandammen Peninsula, Papua

Photo: W. J. Baker

Orania palindan inflorescence, Wandammen Peninsula, Papua

Photo: W. J. Baker

Orania sp. inflorescence, near Tabubil, PNG

Photo: W. J. Baker

Orania sp. with distichous leaves, near Tabubil, PNG

Photo: S. Barrow

Orania regalis, Wandammen Peninsula, Papua

Photo: W. J. Baker

Orania regalis inflorescence, Wandammen Peninsula, Papua

Photo: W. J. Baker

Orania archboldiana inflorescence, Kikori, PNG

Photo: W. J. Baker

Orania archboldiana leaf, Kikori, PNG

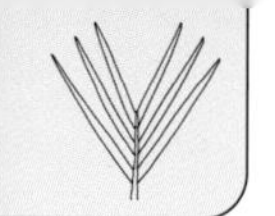

Physokentia

Small – pinnate – crownshaft – no spines – leaflets pointed

Look for:

- Slender, single-stemmed tree palm with stilt roots.
- Inflorescence below the leaves, prophyll not encircling peduncle completely.
- Fruit globose with coat (endocarp) enclosing seed with conspicuous grooves and ridges.

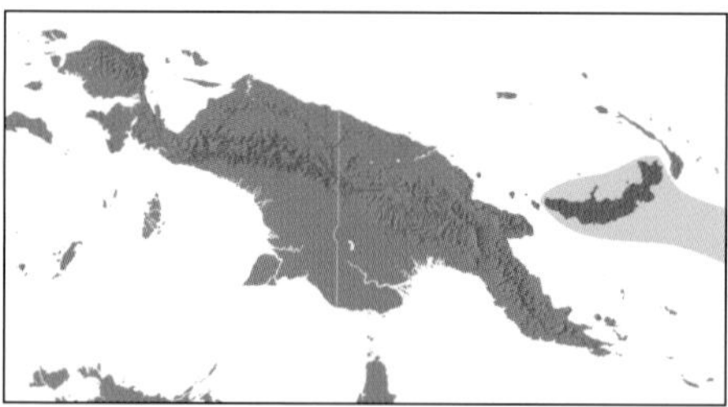

Distribution
Bismarck Archipelago, Solomon Islands, Fiji and Vanuatu, in New Guinea recorded only from New Britain.

Habitat
Submontane to montane rainforest, 450–1800 m.

Number of species
7, of which 1 in New Guinea.

Taxonomic accounts
Moore 1977.

Uses
None recorded.

Habit
Slender, single-stemmed tree palm with stilt roots, height to 15 m, stem diameter to 12 cm, crownshaft conspicuous, monoecious.

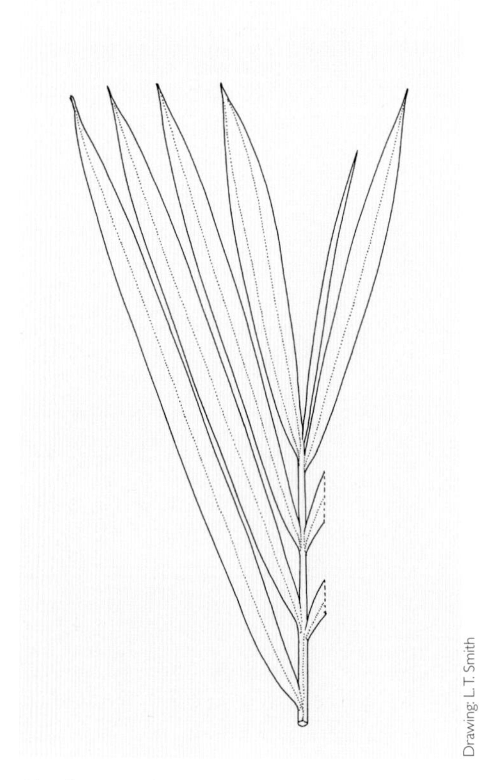

Drawing: L. T. Smith

Physokentia avia leaf tip

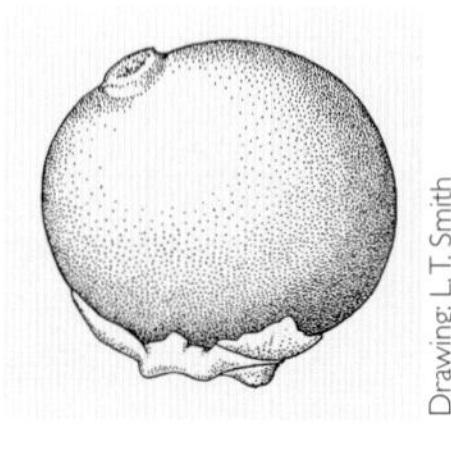

Drawing: L. T. Smith

Physokentia avia fruit

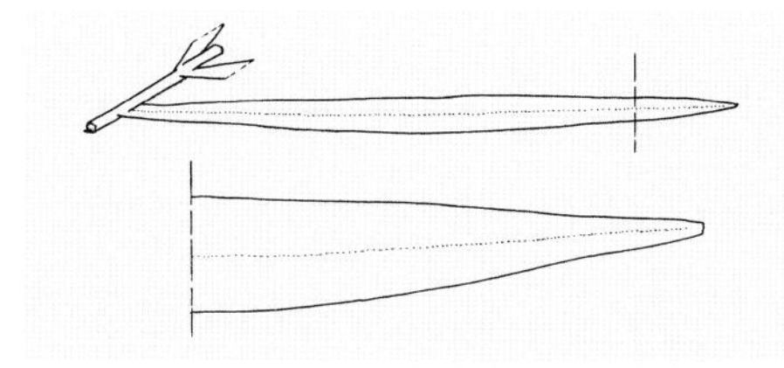

Drawing: P. K. R. Davies

Physokentia leaflet

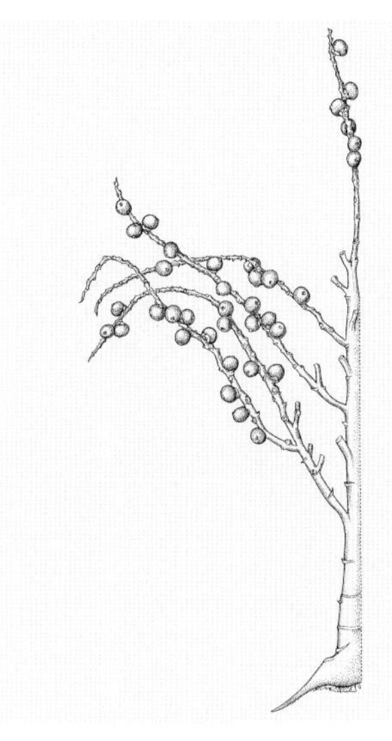

Drawing: L. T. Smith

Physokentia avia inflorescence

Leaf pinnate, 150–200 m long, few in crown, ± arching.

Sheath tubular, c. 80 cm long, forming crownshaft.

Petiole short, c. 20 cm long.

Leaflets c. 26 each side of leaf rachis, to c. 70 cm long, with pointed tips, arranged regularly.

Inflorescence below the leaves, branched to 3 orders, 20–55 cm long.

Prophyll not encircling peduncle completely, peduncular bract much longer than the prophyll, enclosing the inflorescence in bud, dropping off with the prophyll as the inflorescence expands.

Peduncle shorter than inflorescence rachis, 7–9 cm.

Rachillae moderately slender and curved.

Flowers in triads at the base of the branches, pairs of male flowers towards tip, not developing in pits, female flowers much larger than the male flowers.

Fruit black, ± globose, 1.3–1.5 cm diameter, stigmatic remains subapical, flesh thin, endocarp fragile, rounded and pitted.

Seed 1, ± globose, endosperm ruminate.

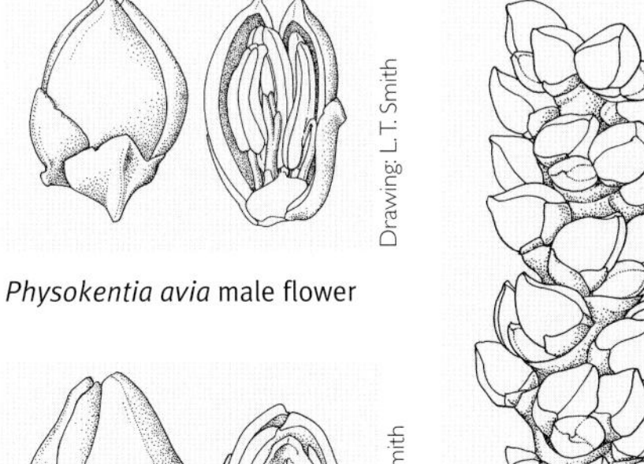

Drawing: L. T. Smith

Physokentia avia male flower

Drawing: L. T. Smith

Physokentia avia female flower

Drawing: L. T. Smith

Physokentia avia flowers on rachilla

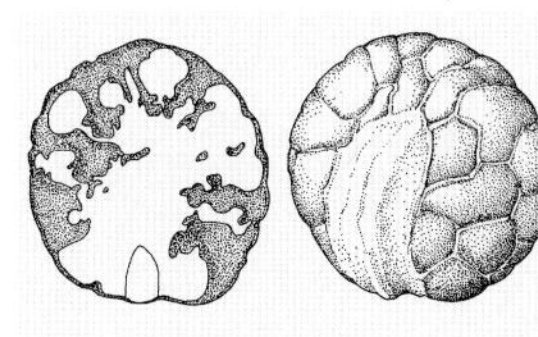

Drawing: L. T. Smith

Physokentia avia seed

Confused with:

- *Areca*: female flowers occurring only at the base of inflorescence branches, no peduncular bract, prophyll completely enclosing inflorescence bud.

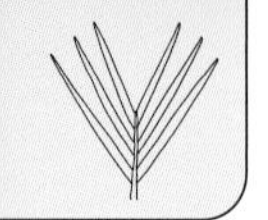

Pigafetta

Tall – pinnate – no crownshaft – spiny – leaflets pointed

Look for:

- Robust, single-stemmed canopy tree palm with arching, spiny leaves, leaf sheaths chalky white.
- Inflorescence between and below the leaves, reaching 6 m long, with long pendulous branches.
- Fruit very small (0.8–1 cm), scales creamy white when ripe.

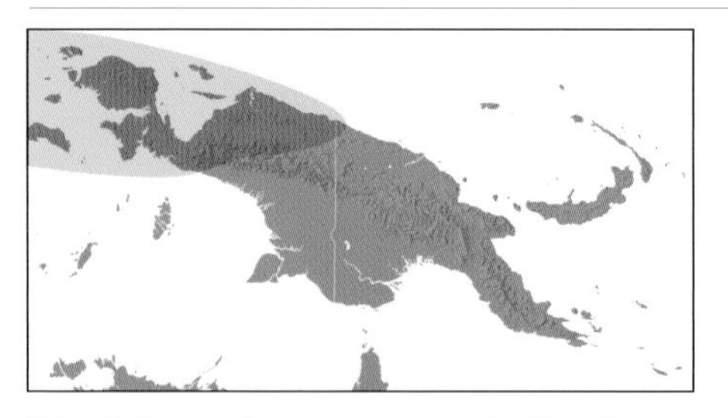

Distribution
Sulawesi to New Guinea, in New Guinea recorded only from the north-west, absent from PNG except for the extreme north-west.

Habitat
Lowland rainforest, sea level to 250 m.

Number of species
2, of which 1 in New Guinea.

Taxonomic accounts
Dransfield 1998.

Uses
Stem for floorboards; leaves for thatch.

Habit
Robust, single-stemmed tree palm, height to 25 m, stem diameter 30–45 cm, stem shiny green, especially in upper parts, no crownshaft, dioecious.

Photo: J. Dransfield

Pigafetta filaris, Andai, Papua

Photo: W. J. Baker

Pigafetta fruit, cultivated, Malaysia

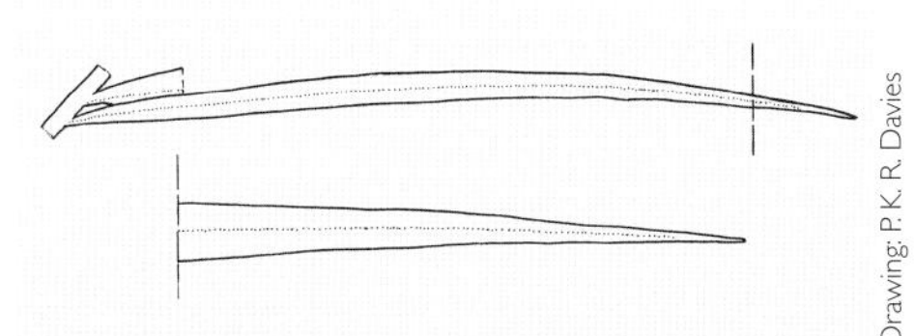

Drawing: P. K. R. Davies

Pigafetta filaris leaflets

Photo: W. J. Baker

Pigafetta filaris stem, private collection, Cairns, Australia

Photo: J. Dransfield

Pigafetta filaris female flowers, Nuni, Papua

Leaf pinnate, 300–600 cm long, 14–20 in crown, armed with spines and bristles, strongly arching.

Sheath splitting to the base opposite the petiole, not forming crownshaft, chalky white.

Petiole absent, though narrow, upper portion of sheath resembles a petiole.

Leaflets c. 60 each side of leaf rachis, to 140 cm long, with pointed tips, arranged regularly, ascending.

Inflorescence between and below the leaves, branched to 2 orders, 150–260 cm long, branches pendulous.

Primary bracts similar, not dropping off as inflorescence expands.

Peduncle shorter than inflorescence rachis, c. 60 cm.

Rachillae moderately slender, pendulous.

Flowers in pairs or solitary throughout the length of the rachilla, not developing in pits.

Fruit creamy white, subglobose, 0.8–1 cm × 0.6–0.7 cm, with vertical rows of shiny scales, stigmatic remains apical, flesh thin.

Seed 1, enclosed in a fleshy coat (sarcotesta), laterally flattened, endosperm homogeneous.

Photo: J. Dransfield

Pigafetta filaris inflorescence, Nuni, Papua

Photo: J. Dransfield

Pigafetta filaris leaf sheath, private collection, Cape Tribulation, Australia

Confused with:

- *Metroxylon*: usually multi-stemmed, stems dying after flowering, fruit large (2.5–3.5 cm).

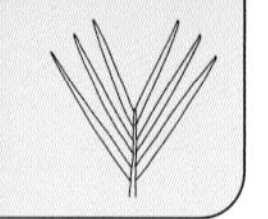

Pinanga

Medium – pinnate – crownshaft – no spines – leaflets lobed or pointed

Look for:

- Moderate, single-stemmed, mid-storey tree palms.
- Inflorescence below leaves, branched to 1 order.
- Female flowers and fruit present from base to tip of inflorescence branches.

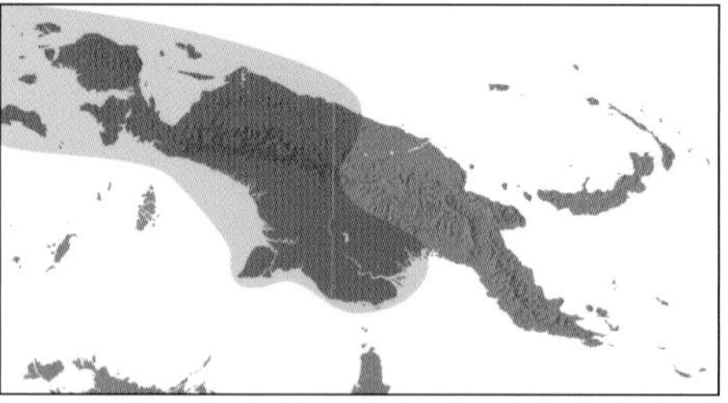

Distribution
South China to India, through South-East Asia to New Guinea, absent from eastern PNG.

Habitat
Lowland to submontane rainforest, sea level to 900 m.

Number of species
Approximately 130, of which 1 in New Guinea.

Taxonomic accounts
Checklist in Govaerts & Dransfield 2005.

Uses
None recorded.

Habit
Moderate, single-stemmed tree palms, height to 15 m, stem diameter 8–12 cm, crownshaft present, monoecious.

Photo: A. S. Barford

Pinanga rumphiana, Bewani, PNG

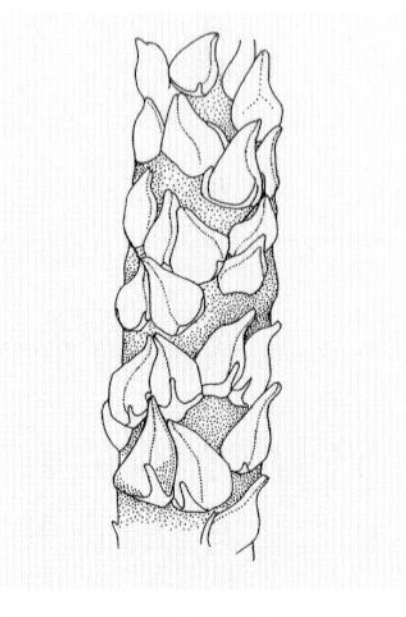

Drawing: L. T. Smith

Pinanga punicea flowers on rachilla

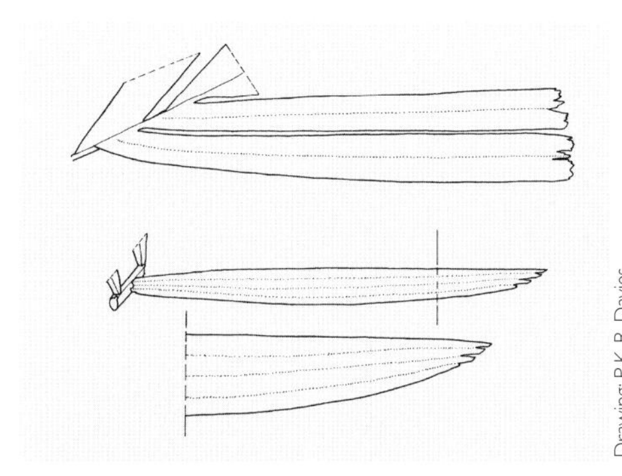
Drawing: P. K. R. Davies

Pinanga leaflets

Photo: S. Barrow

Pinanga rumphiana crown, Wandammen Peninsula, Papua

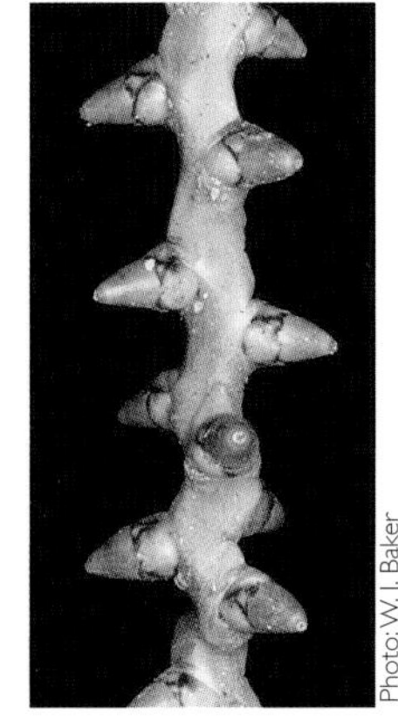
Photo: W.J. Baker

Pinanga rumphiana recently pollinated female flowers, Kikori, PNG

Photo: S. Barrow

Pinanga rumphiana inflorescence in fruit, Wandammen Peninsula, Papua

Leaf pinnate, to 350 cm long, 6–7 in crown, ± arching.

Sheath tubular, to 110 cm long, forming crownshaft.

Petiole short, 10–35 cm long.

Leaflets 29–40 each side of leaf rachis, mostly composed of 2 folds, to 80 cm long, with pointed tips, arranged regularly, apical pair much broader, multifold with lobed tips.

Inflorescence below leaves, branched to 1 order, 35–50 cm long.

Prophyll papery, enclosing the inflorecence in bud, soon dropping off when inflorescence expands, peduncular bract absent.

Peduncle shorter than inflorescence rachis, 6–12 cm.

Rachillae curved at the base then straight, pendulous.

Flowers in triads throughout the length of the rachilla, in a spiral, male flowers fleshy, asymmetrical, much larger than the spherical female flowers.

Fruit red to black, ± ovoid, 1.6–1.7 cm × 1–1.2 cm, stigmatic remains apical, flesh thin, endocarp fibrous.

Seed 1, ± globose, endosperm ruminate.

Confused with:

- *Areca*: female flowers occurring only at the base of inflorescence branches.
- *Hydriastele*: leaflet tips jagged, inflorescence enclosed within prophyll and peduncular bract in bud.

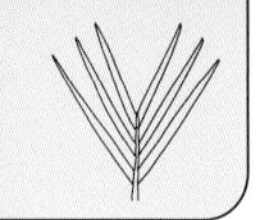

Ptychococcus

Medium to tall – pinnate – crownshaft – no spines – leaflets jagged

Look for:

- Moderate to robust, single-stemmed, mid-storey tree palms, leaves often tilted 90° to one side.
- Inflorescence below the leaves, spreading widely.
- Fruit quite large (4–6 cm), red with black or dark brown, hard coat (endocarp) enclosing seed with conspicuous grooves and ridges.

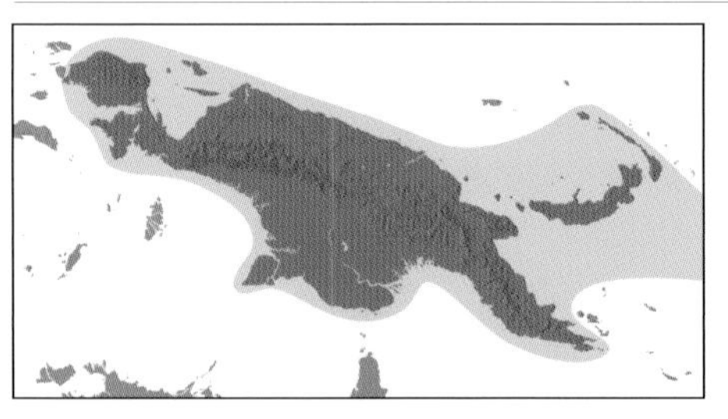

Distribution
Endemic to New Guinea, widespread.

Habitat
Lowland to montane rainforest, sea level to 1800 m.

Number of species
2 species.

Taxonomic accounts
Zona 2005.

Uses
Stem for spears, bows and arrows; sheaths for serving food; shoot apex edible.

Habit
Moderate to robust, single-stemmed tree palms, height to 26 m, stem diameter 9–30 cm, crownshaft present, monoecious.

Photo: W. J. Baker

Ptychococcus paradoxus, Kikori, PNG

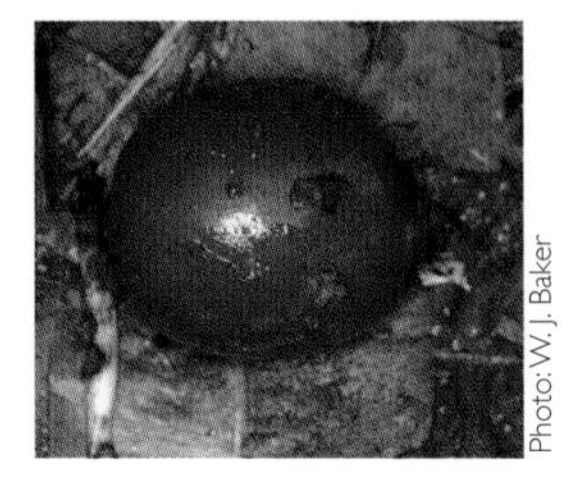

Photo: W. J. Baker

Ptychococcus paradoxus fruit, Wandammen Peninsula, Papua

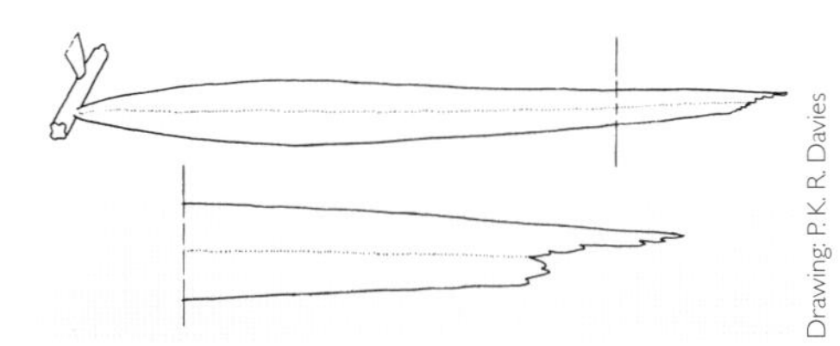

Drawing: P. K. R. Davies

Ptychococcus leaflet

Photo: W. J. Baker

Ptychococcus paradoxus inflorescence, Madang, PNG

Leaf pinnate, 250–650 cm long, 6–13 in crown, slightly arching, typically with some leaves tilted 90° to one side.

Sheath tubular, 60–150 cm long, forming crownshaft.

Petiole 0–150 cm long.

Leaflets 32–95 each side of leaf rachis, to 102 cm long, with jagged tips, arranged regularly, ± horizontal or tilted.

Inflorescence below the leaves, branched to 3 orders, 43–150 cm long, branches widely spreading.

Prophyll and peduncular bract similar, enclosing inflorescence in bud, dropping off as inflorescence expands.

Peduncle shorter than inflorescence rachis, 8–19 cm.

Rachillae quite robust and straight.

Flowers in triads throughout the length of the rachilla, not developing in pits, male flowers larger than female flowers, bullet-shaped.

Fruit red, ovoid, 4–6 cm × 2–3.4 cm, stigmatic remains apical, flesh thick and juicy, endocarp thick, black or dark brown, with conspicuous grooves and ridges.

Seed 1, conforming to endocarp, endosperm ruminate to homogeneous.

Photo: W. J. Baker

Ptychococcus paradoxus flowers, Kikori, PNG

Photo: W. J. Baker

Ptychococcus paradoxus seed in cross section, Wandammen Peninsula, Papua

Confused with:

- *Brassiophoenix*: broad wedge-shaped leaflets, endocarp pale.
- *Drymophloeus*: stilt roots, prophyll usually not dropping off, endocarp pale, lacking ridges.
- *Ptychosperma*: slender palms, fruit small, endocarp pale with shallow ridges and grooves.

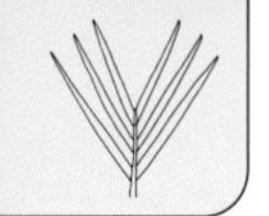

Ptychosperma

Small to medium – pinnate – crownshaft – no spines – leaflets jagged

Look for:

- Slender, single- or multi-stemmed, mid-storey tree palms.
- Inflorescence below the leaves, branches sometimes colourful.
- Fruit red or black, quite small (c. 1 cm), with pale, hard coat (endocarp) enclosing seed, frequently grooved.

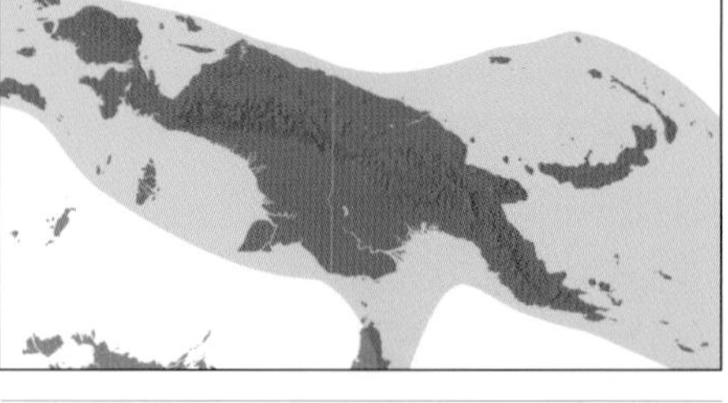
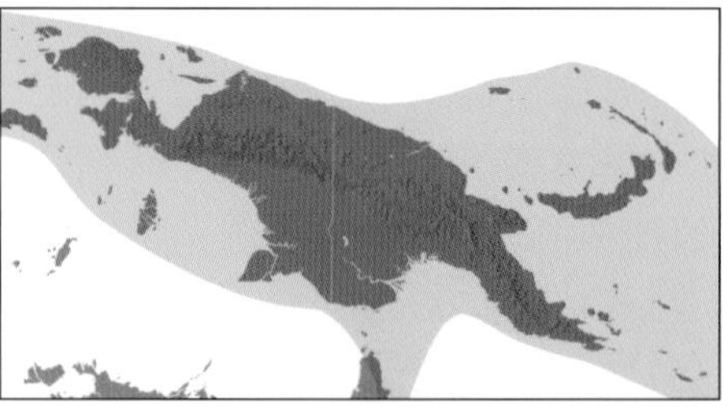

Distribution
New Guinea to Australia and the Solomon Islands, widespread in New Guinea.

Habitat
Lowland to montane rainforest, sea level to 1900 m.

Number of species
Approximately 30, of which 26 in New Guinea.

Taxonomic accounts
Essig 1978.

Uses
Stem for spears, bows and arrows; *P. macarthurii* is one of the most widely cultivated ornamental palms in New Guinea and worldwide.

Habit
Slender or moderately slender, single- or multi-stemmed tree palms, height to 20 m, stem diameter 1.5–10 cm, crownshaft present, monoecious.

Photo: J. Dransfield

Ptychosperma sp., near Ransiki, Papua

Photo: S Zona

Ptychosperma pullenii fruit, cultivated, Fairchild Tropical Botanic Garden, USA

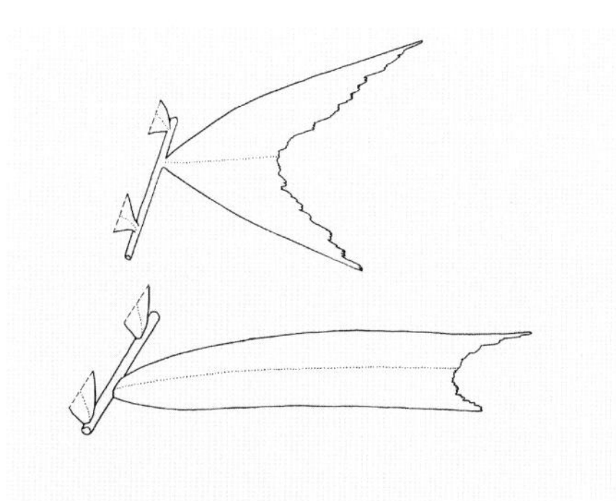

Drawing: P. K. R. Davies

Ptychosperma leaflet variation

Photo: J. Dransfield

Ptychosperma sp. inflorescence, near Ransiki, Papua

Photo: W. J. Baker

Ptychosperma macarthurii male flowers, Forest Research Institute, Malaysia

Photo: S. Zona

Ptychosperma cuneatum leaf, cultivated, Fairchild Tropical Botanic Garden, USA

Leaf pinnate, 44–310 cm long, 5–10 in crown, slightly arching.

Sheath tubular, 18–70 cm long, forming crownshaft.

Petiole 12–61 cm long.

Leaflets 7–47 each side of leaf rachis, to 80 cm long, linear to broadly wedge-shaped, with jagged tips, arranged regularly or irregularly, horizontal.

Inflorescence below the leaves, branched 2–4 orders, 11–90 cm long, branches widely spreading.

Prophyll and peduncular bract similar, enclosing inflorescence in bud, dropping off as inflorescence expands.

Peduncle shorter than inflorescence rachis (rarely longer), 2–7.5 cm.

Rachillae slender to quite robust, often colourful, straight or curving.

Flowers in triads throughout the length of the rachilla, not developing in pits, male flowers larger than female flowers, bullet-shaped.

Fruit red, orange of black, globose to ellipsoid, c. 1 cm × 6–13 cm, stigmatic remains apical, flesh quite thick and juicy, sometimes with irritant needles crystals, endocarp pale, thin, frequently grooved or lobed.

Seed 1, endosperm homogeneous or ruminate.

Confused with:

- *Brassiophoenix*: broad wedge-shaped leaflets, orange, medium-sized fruit (c. 3.5 cm long), endocarp pale with conspicuous grooves and ridges.
- *Dransfieldia*: leaflet tips pointed, prophyll not dropping off.
- *Drymophloeus*: stilt roots, prophyll usually not dropping off.
- *Ptychococcus*: moderate to robust palm, fruit quite large (4–6 cm long), endocarp black or dark brown with conspicuous grooves and ridges.

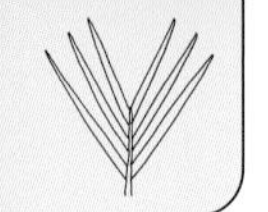

Rhopaloblaste

Tall – pinnate – crownshaft – no spines – leaflets pointed

Look for:

- Robust, single-stemmed, canopy or subcanopy palm with pendulous leaflets.
- Leaf with dark hairs on the leaf rachis, seedling leaf finely pinnate.
- Inflorescence below the leaves, with widely spreading branches and flowers not developing within pits.

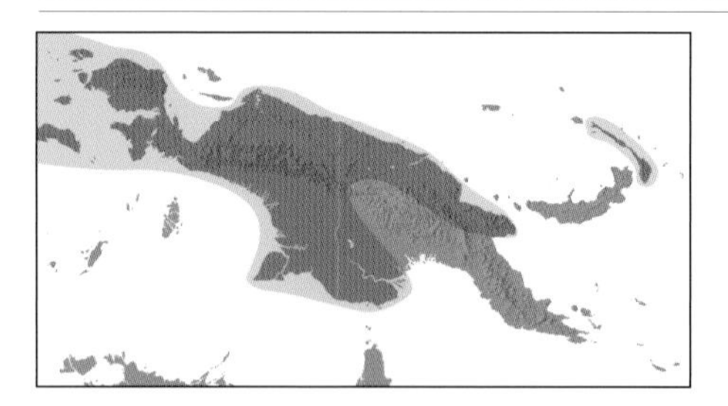

Distribution
Nicobar Islands, Malay Peninsula, Moluccas, New Guinea and Solomon Islands, widespread in New Guinea.

Habitat
Lowland to submontane rainforest, sea level to 900 m.

Number of species
6, of which 3 in New Guinea.

Taxonomic accounts
Banka & Baker 2004.

Uses
Stem for floorboards, bows and arrow tips; shoot apex edible.

Habit
Robust, single-stemmed tree palms, height to 35 m, stem diameter 8–35 cm, crownshaft present, monoecious.

Photo: W. J. Baker

Rhopaloblaste ledermanniana, Wandammen Peninsula, Papua

Photo: W. J. Baker

Rhopaloblaste ledermanniana fruit, Wandammen Peninsula, Papua

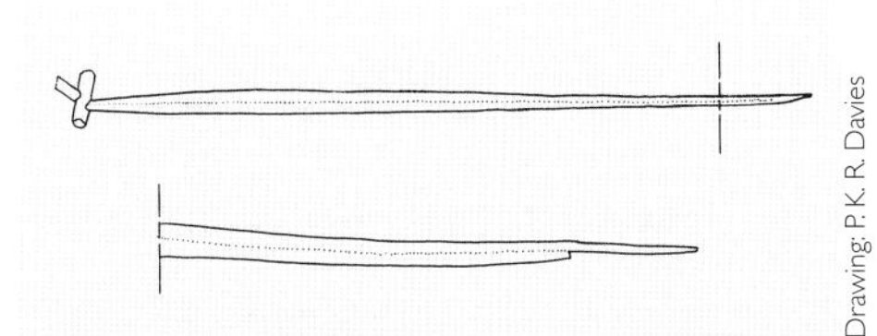
Drawing: P. K. R. Davies

Rhopaloblaste leaflet

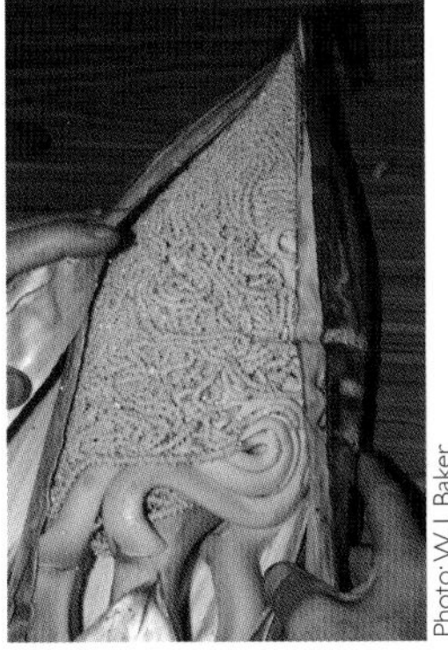
Photo: W. J. Baker

Rhopaloblaste ledermanniana contorted rachillae in inflorescence bud, Timika, Papua

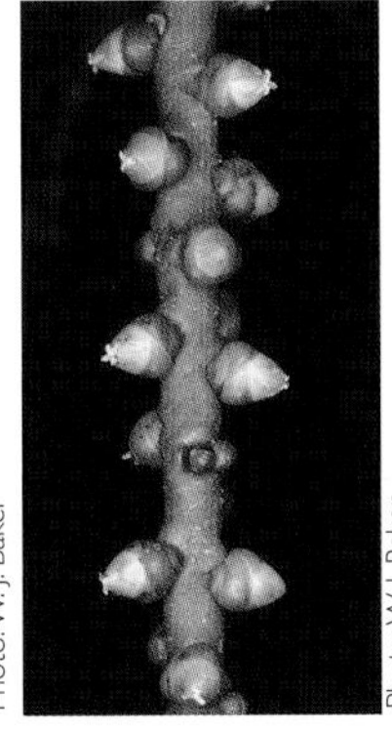
Photo: W. J. Baker

Rhopaloblaste ledermanniana female flowers, Wandammen Peninsula, Papua

Photo: S. Barrow

Rhopaloblaste ledermanniana inflorescence, Wandammen Peninsula, Papua

Leaf pinnate, 200–400 cm long, up to 17 in crown, ± straight, dark hairs on the leaf rachis (most easily observed in recently expanded leaves).

Sheath tubular, 60–150 cm long, forming crownshaft.

Petiole 3–20 cm long.

Leaflets 59–120 each side of leaf rachis, to 112 cm long, with pointed tips, arranged regularly, pendulous.

Seedling leaf finely pinnate.

Inflorescence below the leaves, branched 2–5 orders, 55–130 cm long, branches widely spreading.

Prophyll and peduncular bract similar, enclosing inflorescence in bud, dropping off as inflorescence expands.

Peduncle shorter than inflorescence rachis, 1.5–10 cm.

Rachillae slender and curving, contorted in the inflorescence bud and resembling intestines.

Flowers in triads nearly throughout the length of the rachilla, not developing in pits.

Fruit yellow to red, ellipsoid 1.5–3.5 cm × 0.9–1.8 cm, stigmatic remains apical or slightly to one side of the apex, flesh rather thin and firm, endocarp thin, closely adhering to seed.

Seed 1, globose, endosperm strongly ruminate.

Confused with:

- *Clinostigma*: inflorescence resembling a horse's tail, endosperm homogeneous.
- *Cyrtostachys*: flowers usually developing in pits, fruit very small (0.8–1.2 cm), black, ellipsoid.
- *Dransfieldia*: slender, multi-stemmed (rarely single-stemmed) palm, peduncle same length or longer than inflorescence rachis, prophyll not dropping off.
- *Heterospathe*: no crownshaft, leaf sheaths fibrous, prophyll not dropping off.
- *Hydriastele*: leaflet tips jagged (except *H. costata*), inflorescence resembling a horse's tail.

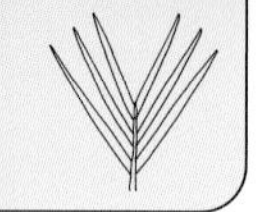

Sommieria

Small – pinnate – no crownshaft – no spines – leaflets toothed or pointed

Look for:

- Slender, single-stemmed, understorey palm with fibrous leaf sheath and leaf blade chalky white below.
- Inflorescence between the leaves, with few branches, inflorescence below the leaves, and flowers developing within pits.
- Fruit small (1 cm long) with corky warts, pink when ripe.

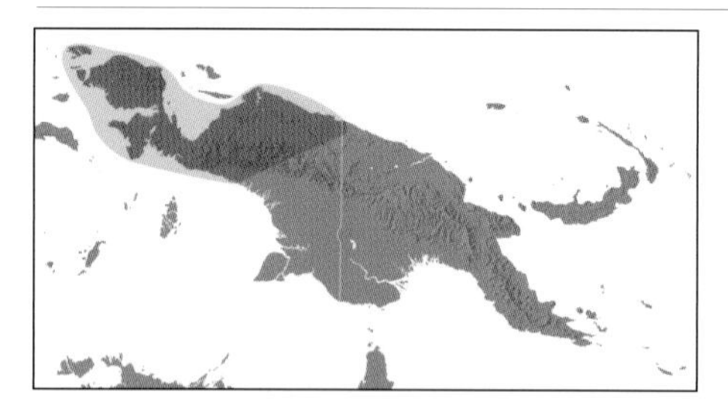

Distribution
Endemic to New Guinea, widespread in western New Guinea, only few records from western PNG.

Habitat
Lowland rainforest, sea level to 500 m.

Number of species
1 species.

Taxonomic accounts
Heatubun 2002, Stauffer et al. 2004.

Uses
Leaves for thatch and wrapping food.

Habit
Slender, single-stemmed palm, height to 3 m, sometimes stemless, stem diameter 3–4 cm, crownshaft absent, monoecious.

Photo: J. Dransfield

Sommieria leucophylla, Timika, Papua

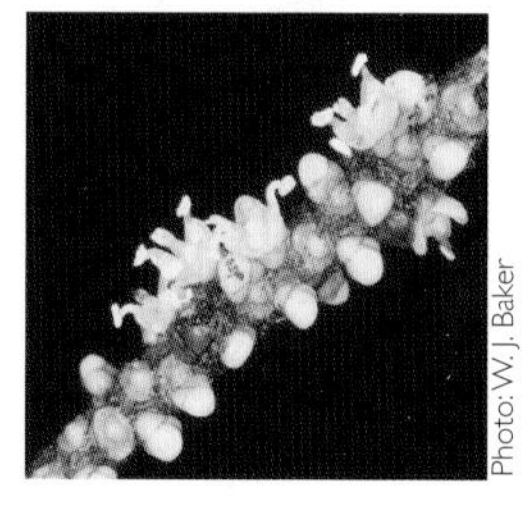

Photo: W. J. Baker

Sommieria leucophylla male flowers, Wandammen Peninsula, Papua

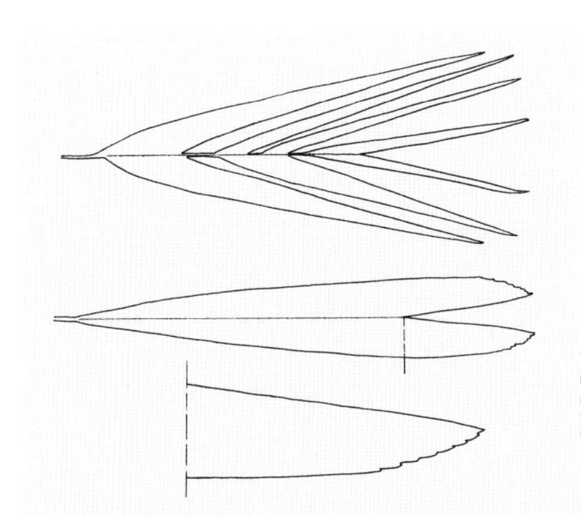
Drawing: P. K. R. Davies

Sommieria leaf variation

Leaf entire or pinnate with few broad leathery leaflets, 180 cm long, up to 40 in crown, ± arching.

Sheath splitting ± to base opposite the petiole, to 30 cm long, not forming a crownshaft.

Petiole short, 10–40 cm long.

Leaflets (where blade divided) 2–3 on each side of leaf rachis, multifold, 50–180 cm long, with pointed or toothed tips, upper surface dark green, lower surface chalky white.

Inflorescence between the leaves, branched to 1 order with few branches, to 160 cm long, flowers developing within pits.

Prophyll hidden among leaf sheaths, peduncular bract enclosing the inflorescence in bud, borne at the tip of the peduncle, splitting and usually dropping as the inflorescence expands.

Peduncle longer than inflorescence rachis, to 150 cm.

Rachillae slender and straight.

Flowers in triads throughout the length of the rachilla, developing in shallow pits.

Fruit yellow-brown to pink, round, up to 1 cm diameter, the surface cracked into corky warts, flesh soft, endocarp smooth.

Seed 1, round, endosperm homogeneous.

Photo: J. Dransfield

Sommieria leucophylla inflorescence in fruit, Timika, Papua

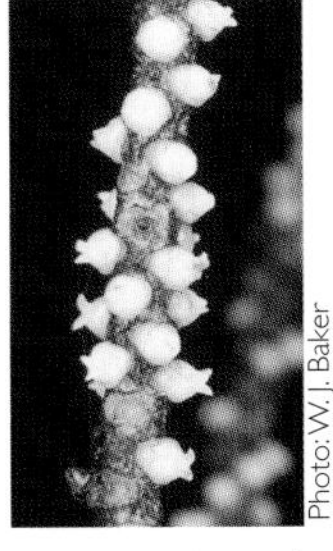
Photo: W. J. Baker

Sommieria leucophylla female flowers, Wandammen Peninsula, Papua

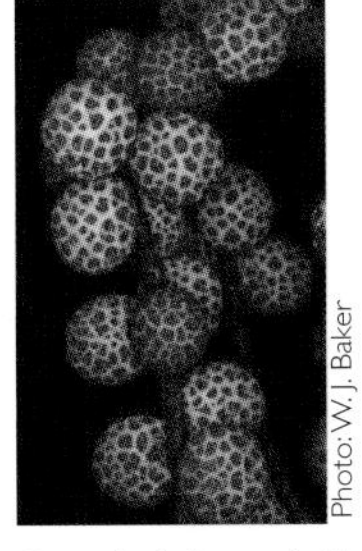
Photo: W. J. Baker

Sommieria leucophylla fruit, Timika, Papua

Confused with:

- *Calyptrocalyx*: leaves not chalky white beneath, inflorescence spicate.
- *Korthalsia* (juveniles only): leaves spiny.

Checklist of New Guinea palm genera and synonyms

Actinorhytis *H.Wendl. & Drude*

Areca *L.*
Mischophloeus Scheff.
Gigliolia Becc.
Pichisermollia H.Mont.-Neto

Arenga *Labill.*
Saguerus Steck
Gomutus Correa
Blancoa Blume
Didymosperma H.Wendl. & Drude

Borassus *L.*
Lontarus Adans.

Brassiophoenix *Burret*

Calamus *L.*
Rotanga Boehm.
Rotang Adans.
Palmijuncus Kuntze
Calospatha Becc.
Zalaccella Becc.
Schizospatha Furtado
Cornera Furtado

Calyptrocalyx *Blume*
Paralinospadix Burret

Caryota *L.*
Schunda-Pana Adans.
Thuessinkia Korth. ex Miq.

Clinostigma *H.Wendl.*
Exorrhiza Becc.
Bentinckiopsis Becc.
Clinostigmopsis Becc.

Cocos *L.*

Corypha L.
Codda-Pana Adans.
Taliera Mart.
Gembanga Blume

Cyrtostachys *Blume*

Daemonorops *Blume*

Dransfieldia *Baker et al.*

Drymophloeus *Zipp.*
Coleospadix Becc.
Rehderophoenix Burret
Saguaster Kuntze

Heterospathe *Scheff.*
Ptychandra Scheff.
Barkerwebbia Becc.
Alsmithia H.E.Moore

Hydriastele *H.Wendl. & Drude*
Adelonenga Hook.f.
Gronophyllum Scheff.
Gulubia Becc.
Gulubiopsis Becc.
Kentia Blume
Leptophoenix Becc.
Nengella Becc.
Paragulubia Burret
Siphokentia Burret

Korthalsia *Blume*
Calamosagus Griff.

Licuala *Thunb.*
Pericycla Blume
Dammera Lauterb. & K.Schum.

Linospadix H.Wendl.
Bacularia F.Muell. ex Hook.f.

Livistona *R.Br.*
Saribus Blume
Wissmannia Burret

Metroxylon *Rottb.*
Sagus Steck
Coelococcus H.Wendl.

Nypa *Steck*
Nipa Thunb.

Orania *Zipp.* ex *Blume*
Arausiaca Blume
Macrocladus Griff.
Sindroa Jum.
Halmoorea J.Dransf. & N.W.Uhl

Physokentia *Becc.*
Goniosperma Burret
Goniocladus Burret

Pigafetta *(Blume) Becc.*

Pinanga *Blume*
Cladosperma Griff.
Ophiria Becc.
Pseudopinanga Burret

Ptychococcus *Becc.*

Ptychosperma *Labill.*
Seaforthia R.Br.
Actinophloeus (Becc.) Becc.
Romanowia Sander ex André
Strongylocaryum Burret

Rhopaloblaste *Scheff.*
Ptychoraphis Becc.

Sommieria *Becc.*

Further reading

Bachman, S., W.J. Baker, N. Brummitt, J. Dransfield & J. Moat. 2004. Elevational gradients, area and tropical island diversity: an example from the palms of New Guinea. *Ecography* 27: 299–310.

Baker, W.J. 2002a. The Palms of New Guinea Project. *Flora Malesiana Bulletin* 13: 35–37.

Baker, W.J. 2002b. Two unusual *Calamus* species from New Guinea. *Kew Bulletin* 57: 719–724.

Baker, W.J., R.P. Bayton, J. Dransfield & R.A. Maturbongs. 2003. A revision of the *Calamus aruensis* (Arecaceae) complex in New Guinea and the Pacific. *Kew Bulletin* 58: 351–370.

Baker, W.J. & J. Dransfield. 2002a. *Calamus maturbongsii*, an unusual new rattan species from New Guinea. *Kew Bulletin* 57: 725–728.

Baker, W.J. & J. Dransfield. 2002b. *Calamus longipinna (Arecaceae: Calamoideae)* and its relatives in New Guinea. *Kew Bulletin* 57: 853–866.

Baker, W.J. & J. Dransfield. 2006. Arecaceae of Papua. In: A.J. Marshall & B.M. Beehler (eds.), *The Ecology of Papua,* in press. Periplus Editions, Singapore.

Baker, W.J. & A.H.B. Loo. 2004. A synopsis of the genus *Hydriastele* (Arecaceae). *Kew Bulletin* 59: 61–68.

Baker, W.J., S. Zona, C.D. Heatubun, C.E. Lewis, R.A. Maturbongs & M.V. Norup. 2006. *Dransfieldia* (Arecaceae) – a new palm genus from western New Guinea. *Systematic Botany* 31: 61–69.

Banka, R. & W.J. Baker. 2004. A monograph of the genus *Rhopaloblaste* (Arecaceae). *Kew Bulletin* 59: 47–60.

Banka, R. & A.S. Barfod. 2004. A new, spectacular species of *Licuala* (Arecaceae, Coryphoideae) from New Guinea. *Kew Bulletin* 59: 73–75.

Barfod, A.S. 2000. A new species of *Licuala* from New Guinea. *Palms* 44: 198–201.

Barfod, A.S., R. Banka & J.L. Dowe. 2001. *Field Guide to Palms in Papua New Guinea.* AAU Reports 40, Department of Systematic Botany, University of Aarhus.

Bayton, R.P. 2005. Borassus *L. and the Borassoid Palms: Systematics and Evolution.* PhD Thesis, University of Reading.

Dowe, J.L. 2001. *Studies in the Genus* Livistona *(Coryphoideae: Arecaceae)*. PhD Thesis, James Cook University, Townsville.

Dowe, J. & A.S. Barfod. 2001. New species of *Livistona* R. Br. (Arecaceae) from north Queensland and New Guinea. *Austrobaileya* 6: 165–174.

Dowe, J.L. & M.D. Ferrero. 2001. Revision of *Calyptrocalyx* and the New Guinea species of *Linospadix* (Linospadicinae: Arecoideae: Arecaceae). *Blumea* 46: 207–251.

Dransfield, J. 1981. A synopsis of *Korthalsia* (Palmae–Lepidocaryoideae). *Kew Bulletin* 36: 163–194.

Dransfield, J. 1982. *Clinostigma* in New Ireland. *Principes* 26: 73–76.

Dransfield, J. 1986. A guide to collecting palms. *Annals of the Missouri Botanical Garden* 73: 166–176.

Dransfield, J. 1998. *Pigafetta. Principes* 42: 34–40.

Dransfield, J. & W.J. Baker. 2003. An account of the Papuasian species of *Calamus* (Arecaceae) with paired fruit. *Kew Bulletin* 58: 371–387.

Dransfield, J. & H. Beentje. 1996. *Lexicon Palmarum.* Editions Champflour, Marly-le-Roi.

Dransfield, J., G.G. Hambali, R.A. Maturbongs & C.D. Heatubun. 2000. *Caryota zebrina. Palms* 44: 170–174.

Dransfield, J. & N.W. Uhl. 1998. Palmae. In: K. Kubitzki (ed.), *The Families and Genera of Vascular Plants,* volume 4, pp. 306–389. Springer, Berlin.

Dransfield, J., N.W. Uhl, C.B. Asmussen, W.J. Baker, M.M. Harley & C.E. Lewis. 2006. A new phylogenetic classification of the palm family, *Arecaceae. Kew Bulletin* 60: 559–569

Essig, F.B. 1977. A preliminary analysis of the palm flora of New Guinea and the Bismarck Archipelago. *Papua New Guinea Botany Bulletin* 9.

Essig, F.B. 1978. A revision of the genus *Ptychosperma* Labill. (Arecaceae). *Allertonia* 7: 415–478.

Essig, F.B. 1980. The genus *Orania* in New Guinea. *Lyonia* 1: 211–233.

Essig, F.B. 1995. A checklist of the palms of the Bismarck Archipelago. *Principes* 39: 123–129.

Evans, T.D., K. Sengdala, O.V. Viengkhan & B. Thammavong. 2001. *A Field Guide to the Rattans of Lao PDR.* Royal Botanic Gardens, Kew.

Ferrero, M.D. 1997. A checklist of Palmae for New Guinea. *Palms & Cycads* 55/56: 2–39.

Flynn, T. 2004. *Morphological Variation and Species Limits in the Genus* Areca *(Palmae) in New Guinea and the Solomon Islands*. MSc Thesis, University of Wales, Bangor.
Govaerts, R. & J. Dransfield. 2005. *World Checklist of Palms*. Royal Botanic Gardens, Kew.
Harries, H.C. 1978. Evolution, dissemination and classification of Cocos nucifera L. *Botanical Review* 44: 265–319
Hay, A.J. 1984. Palmae. In: R.J. Johns & A.J. Hay (eds.), *A Guide to the Monocotyledons of Papua New Guinea*, volume 3.
Heatubun, C.D. 2002. A monograph of *Sommieria* (Arecaceae). *Kew Bulletin* 57: 599–611.
Heatubun, C.D. 2005. *A Monograph of the Palm Genus* Cyrtostachys *Blume* (*Arecaceae*). Masters Thesis, Institut Pertanian Bogor, Bogor, Indonesia.
Jones, D.L. 1995. *Palms Throughout the World*. Reed, Chatswood, Australia.
Keim, A.P. 2003. *A Monograph of the Genus* Orania *Zippelius (Arecaceae: Oraniinae)*. PhD Thesis, University of Reading.
Kjær, A., A.S. Barfod, C.B. Asmussen & O. Seberg. 2004. Investigation of genetic and morphological variation in the Sago palm (*Metroxylon sagu*; Arecaceae) in Papua New Guinea. *Annals of Botany* 94: 109–117.
Loo, A.H.B., J. Dransfield, M.W. Chase & W.J. Baker. 2006. Low copy nuclear DNA, phylogeny and the evolution of dichogamy in the betel nut palms and their relatives (Arecinae; Arecaceae). *Molecular Phylogenetics and Evolution* 39: 598–618.
Maturbongs, R.A. 2003. Daemonorops *East of Wallace's line*. Masters Thesis, Universitas Indonesia, Depok.
Mogea, J.P. 1991. *Revisi Marga* Arenga *(Palmae)*. PhD Thesis, Universitas Indonesia.
Moore, H.E. 1969. New palms from the Pacific, III. *Principes* 13: 99–108.
Moore, H.E. 1977. New palms from the Pacific, IV. *Principes* 21: 86–88.
Rauwerdink, J.B. 1986. An essay on *Metroxylon*, the Sago palm. *Principes* 30: 165–180.
Stauffer, F.W, W.J. Baker, J. Dransfield & P. K. Endress. 2004. Comparative floral structure and systematics of *Pelagodoxa* and *Sommieria* (Arecaceae). *Botanical Journal of the Linnean Society* 146: 27–39.
Tomlinson, P.B. 1990. *The Structural Biology of Palms*. Clarendon Press, Oxford.
Uhl, N.W., Dransfield, J. 1987. *Genera Palmarum: a classification of palms based on the work of H.E. Moore, Jr.* International Palm Society and L.H. Bailey Hortorium, Lawrence, KS.
Zona, S. 1999. Revision of *Drymophloeus* (Arecaceae: Arecoideae). *Blumea* 44: 1–24.
Zona, S. 2005. A revision of *Ptychococcus* (Arecaceae). *Systematic Botany* 30: 520–529.
Zona, S. & F.B. Essig. 1999. How many species of *Brassiophoenix*? *Palms* 43: 45–48.

Index of palm terminology

Taxonomic index

(**Bold** numbers indicate the main account for each genus)